Pharmacogenomics in Psychiatry

Advances in Biological Psychiatry

Vol. 25

Pharmacogenomics in Psychiatry

Volume Editors

M. Schwab Stuttgart/Tübingen
W.P. Kaschka Ulm/Ravensburg
E. Spina Messina

8 figures, 4 in color, and 17 tables, 2010

KARGER Basel · Freiburg · Paris · London · New York · Bangalore · Bangkok · Shanghai · Singapore · Tokyo · Sydney

Prof. Dr. Matthias Schwab
Dr. Margarethe Fischer-Bosch
Institute of Clinical Pharmacology
Auerbachstrasse 112
DE–70376 Stuttgart (Germany)
and
Department of Clinical
Pharmacology
Institute of Experimental and
Clinical Pharmacology and
Toxicology, University Hospital,
Tübingen, Germany

Prof. Dr. Wolfgang P. Kaschka
Department of Psychiatry I
University of Ulm
ZfP Südwürttemberg
Weingartshofer Strasse 2
DE–88214 Ravensburg (Germany)

Prof. Dr. Edoardo Spina
Department of Clinical and
Experimental Medicine and
Pharmacology
Section of Pharmacology
University of Messina
Policlinico Universitario
Via Consolare Valeria
IT–98125 Messina (Italy)

Library of Congress Cataloging-in-Publication Data
Pharmacogenomics in psychiatry / volume editors, M. Schwab, W.P. Kaschka, E. Spina.
p. ; cm. -- (Advances in biological psychiatry, ISSN 0378-7354 ; v. 25)
Includes bibliographical references and indexes.
ISBN 978-3-8055-9498-1 (hard cover : alk. paper) -- ISBN 978-3-8055-9499-8 (e-ISBN)
1. Mental illness--Chemotherapy. 2. Mental illness--Genetic aspects. 3. Pharmacogenomics. I. Schwab, M. (Matthias) II. Kaschka, Wolfgang P. III. Spina, E. (Edoardo) IV. Series: Advances in biological psychiatry, v. 25. 0378-7354 ;
[DNLM: 1. Mental Disorders--drug therapy. 2. Mental Disorders--genetics. 3. Pharmacogenetics--methods. 4. Psychotropic Drugs--pharmacokinetics. W1 AD44 v.25 2010 / WM 402 P5345 2010]
RC483.P4916 2010
616.89'061--dc22
2010017163

Bibliographic Indices. This publication is listed in bibliographic services, including Pub Med/MEDLINE.

Disclaimer. The statements, opinions and data contained in this publication are solely those of the individual authors and contributors and not of the publisher and the editor(s). The appearance of advertisements in the book is not a warranty, endorsement, or approval of the products or services advertised or of their effectiveness, quality or safety. The publisher and the editor(s) disclaim responsibility for any injury to persons or property resulting from any ideas, methods, instructions or products referred to in the content or advertisements.

Drug Dosage. The authors and the publisher have exerted every effort to ensure that drug selection and dosage set forth in this text are in accord with current recommendations and practice at the time of publication. However, in view of ongoing research, changes in government regulations, and the constant flow of information relating to drug therapy and drug reactions, the reader is urged to check the package insert for each drug for any change in indications and dosage and for added warnings and precautions. This is particularly important when the recommended agent is a new and/or infrequently employed drug.

www.karger.com
Printed in Switzerland on acid-free and non-aging paper (ISO 9706) by Reinhardt Druck, Basel
ISSN 0378–7354
ISBN 978–3–8055–9498–1
e-ISBN 978–3–8055–9499–8

Contents

Preface

Psychiatrists have followed developments in the rapidly expanding field of pharmacogenetics and pharmacogenomics with great interest. Methods for making drug treatment more effective have been the central focus in psychiatric medicine in recent years ('the right drug for the right patient'). However, improvements in drug efficacy and tolerability and finding of the optimal dosage can only be realized if in vivo mechanisms of drug action (pharmacodynamics) and ADME (absorption, distribution, metabolism, excretion) processes (pharmacokinetics) of psychopharmacological agents are better understood. The urgent need for further progress in this field is obvious.

A number of comprehensive multicenter studies have shown that, in terms of efficacy and tolerability, the pharmacological treatment strategies presently available for common psychiatric diseases are still far from satisfactory. For example, in the first level of the Sequenced Treatment Alternatives to Relieve Depression (STAR*D) trial, only about 30% of patients were in remission after follow-up of 12 weeks' drug therapy using a selective serotonin reuptake inhibitor [1]. There is also a substantial body of evidence available indicating that even amongst antidepressant responders residual symptoms are common and associated with poorer psychosocial functioning as well as increased relapse rates [2]. As far as schizophrenia is concerned, pharmacological treatments which block the dopamine system are usually effective for delusions and hallucinations, but less so for disabling cognitive and motivational impairments [3, 4]. The Clinical Antipsychotic Trials of Intervention Effectiveness (CATIE) study showed, among other things, that non-compliance to prescribed medication is a major clinical problem in antipsychotic therapy. For instance, Lieberman et al. [5] reported that the majority of patients in each of several study groups receiving different antipsychotics discontinued their assigned treatment owing to inefficacy or intolerable side effects or for other reasons.

In this volume of *Advances in Biological Psychiatry*, current progress and perspectives in pharmacogenetic testing of drug-metabolizing enzymes, drug transporters and other drug targets involved in the response to psychotropic agents are described extensively. There are great expectations that in the near future pharmacogenomics will provide us with the means of identifying subgroups of patients which are at risk of therapeutic failure or more vulnerable to certain adverse effects of psychopharmacological agents. To mention just two examples from recent years: an association has been detected between treatment-emergent suicidal ideation in individuals receiving citalopram therapy and polymorphisms near the cyclic adenosine monophosphate response element binding protein (CREB-1) gene [6]. In another example, data from a study conducted by Opgen-Rhein and Dettling [7] suggest that certain groups of patients carry a genetically determined proneness to clozapine-induced agranulocytosis (as described in this volume by Buckley et al.). Of course, these findings have to be replicated and further research will be necessary in these areas.

Although polymorphisms of genes are useful markers to explain interindividual variability in ADME processes and drug response, the early enthusiasm about the promise of individualized therapy in psychiatric diseases and personalized medicine in general has been tempered by the complexity and multifactorial character of drug responses [8]. Impressive developments in a number of genomic profiling approaches – such as microarray technologies, genome-wide association studies and most recently the next-generation sequencing technique – have given rise to the hope that more comprehensive information about a patient's genomic profile will be available in the near future for improvements in patient care in psychiatric medicine. Moreover, epigenetic mechanisms (i.e. DNA methylation, histone modification and regulation by miRNA), which may result in individual modification of gene expression and phenotype without affecting the DNA sequence, need to be considered as important players in the complex interactions between the multiple genes and environmental factors shaping a distinct phenotype.

This volume presents a timely overview of what has been achieved up to now in the field of psychiatric pharmacogenomics and some promising directions and perspectives for future research that could ultimately lead to substantial improvements in treatment.

We gratefully acknowledge the continuous support and advice of Gunhild Wolf-Hinderling and Gabriella Karger of Karger Publishers, Basel, during the preparation of this book.

Matthias Schwab, Stuttgart/Tübingen
Wolfgang P. Kaschka, Ulm/Ravensburg
Edoardo Spina, Messina

References

1 Trivedi MH, Rush AJ, Wisniewski SR, Nierenberg AA, Warden D, Ritz L, Norquist G, Howland RH, Lebowitz B, McGrath PJ, Shores-Wilson K, Biggs MM, Balasubramani GK, Fava M; STAR*D Study Team: Evaluation of outcomes with citalopram for depression using measurement-based care in STAR*D: implications for clinical practice. Am J Psychiatry 2006;163:28–40.

2 Papakostas GI, Fava M, Thase ME: Treatment of SSRI-resistant depression: a meta-analysis comparing within- versus across-class switches. Biol Psychiatry 2008;63:699–704.

3 Keefe RS, Bilder RM, Davis SM, Harvey PD, Palmer BW, Gold JM, Meltzer HY, Green MF, Capuano G, Stroup TS, McEvoy JP, Swartz MS, Rosenheck RA, Perkins DO, Davis CE, Hsiao JK, Lieberman JA; CATIE Investigators; Neurocognitive Working Group: Neurocognitive effects of antipsychotic medications in patients with chronic schizophrenia in the CATIE trial. Arch Gen Psychiatry 2007;64:633–647.

4 Van Os J, Kapur S: Schizophrenia. Lancet 2009;374:635–645.

5 Lieberman JA, Stroup TS, McEvoy JP, Swartz MS, Rosenheck RA, Perkins DO, Keefe RSE, Davis SM, Davis CE, Lebowitz BD, Severe J, Hsiao JK; Clinical Antipsychotic Trials of Intervention Effectiveness (CATIE) Investigators: Effectiveness of antipsychotic drugs in patients with chronic schizophrenia. N Engl J Med 2005;353:1209–1223.

6 Perlis RH, Purcell S, Fava M, Fagerness J, Rush JR, Trivedi MH, Smoller JW: Association between treatment-emergent suicidal ideation with citalopram and polymorphisms near cyclic adenosine monophosphate response element binding protein in the STAR*D study. Arch Gen Psychiatry 2007;64:689–697.

7 Opgen-Rhein C, Dettling M: Clozapine-induced agranulocytosis and its genetic determinants. Pharmacogenomics 2008;9:1101–1111.

8 Kirchheiner J, Schwab M: Heterogeneity of drug responses and individualization of therapy; Waldman SA, Terzic A: Pharmacology and Therapeutics: Principles to Practice. Philadelphia, Elsevier, 2008, pp 225–238.

Schwab M, Kaschka WP, Spina E (eds): Pharmacogenomics in Psychiatry.
Adv Biol Psychiatry. Basel, Karger, 2010, vol 25, pp 1–11

One Tablet or Two? Towards the Development of Pharmacogenomically Informed Drug Dose Individualization

Julia Kirchheiner[a] · Angela Seeringer[a] · Roberto Viviani[b] · Steve Hodgkinson[c]

[a]Institute of Pharmacology of Natural Products and Clinical Pharmacology and Departments of [b]Psychiatry and Psychotherapy III and [c]Psychiatry and Psychotherapy I, University of Ulm, Ulm, Germany

Abstract

Over 50 years of pharmacogenetic research have produced many examples of how inherited variability in drug metabolism can influence individual responses to psychotropic drugs. However, this knowledge has largely failed to be translated into broadly applicable strategies for improvements in individual drug treatment in psychiatry. One important argument brought against the widespread adoption of pharmacogenetics as a clinical tool is the lack of available evidence showing its influence on contemporary clinical praxis and its potential role in improving the risk/benefit ratio for the patients. Individual drug-metabolizing capacity is assessed by genotyping drug-metabolizing enzymes. The information gained from genotyping patients may be used to adjust initial and maintenance drug doses according to genotype. However, even where the consequences of genotype on pharmacokinetics are significant and well-known, as in the case of many tricyclic antidepressants and several selective serotonin reuptake inhibitors, there is still considerable controversy as to whether the adjustment of dosage (based upon genetic information) to improve therapeutic efficacy and/or to reduce the occurrence of adverse events is of any practical importance in clinical practice. Pharmacogenetic studies can improve our understanding of the functional consequence of a genetic variant in the clinical setting, and the use of intermediate phenotypes instead of broad outcome parameters (such as drug response or remission) might improve our knowledge regarding the clinical response of an individual with a specific genotype to a specific drug. Here, we review the potential impact of adopting an integrated approach to patient treatment, which combines the use of intermediate phenotypes that arise from common genetic polymorphisms in drug metabolism enzymes, the monitoring of the therapy progress, and the possibility of pharmacogenetic-based response prediction in depression.

Need for Individualized Therapy in Psychiatry

The response of individual patients given the same dose of the same drug varies considerably. Some patients exhibit the desired clinical improvements associated with a particular drug, whilst others exhibit little or no improvement. A proportion of these patients may suffer from well-documented adverse drug reactions and, very rarely, individual patients will die as a result of these side effects. In the field of psychiatric drug treatment, it is particularly difficult for physicians to prescribe the optimal drug in the optimal dose for each patient, since the prediction of a patient's response to any specific drug is seldom possible.

To understand why individual patients respond differently to drugs, it is useful to consider the progress of a particular drug from its initial administration to its observed clinical effect. The clinical effect of a drug is dependent on: (1) its systemic concentration and (2) its concentration at the drug target site. The systemic concentration of a drug depends on a number of pharmacokinetic factors such as age, body mass index and sex. The potential influence of these factors should be taken into account when determining the dosage for an individual patient. In addition, hereditary variation in drug-metabolizing enzymes and drug transporters may also exert considerable influence on drug concentrations. Many enzymes involved in drug metabolism carry genetic variants (polymorphisms) which can decrease enzyme activity or, in certain cases, even lead to the complete inactivation of the enzyme in question [1]. Heterozygous carriers of drug-metabolizing enzyme genetic polymorphisms have enzyme activity levels intermediate to both homozygous and wild-type carriers. Such variation can lead to differences in pharmacokinetics of the drug, such as higher or lower blood or tissue concentrations of a drug and its metabolites during the so-called phase I reactions. These initial steps in the biotransformation of the drug are small molecular modifications, such as oxidations and reductions, and are mostly mediated by the cytochrome P450 (CYP) enzyme family. For example, the CYP2D6 enzyme is known to be involved in the metabolism of many psychotropic drugs, such as antidepressants and antipsychotics. CYP2D6 has been found to exhibit high individual variability in catalytic activity mainly caused by genetic polymorphisms [2]. The phenotype determined by *CYP2D6* genotype arises from a number of functional *CYP2D6* alleles, so the presence of two, one or no functional *CYP2D6* gene copies results in rapid or extensive metabolizer, intermediate metabolizer and slow or poor metabolizer phenotypes, respectively [3, 4] (www.cypalleles.ki.se). Furthermore, inheritance of three or more functional alleles by gene duplication gives rise to the ultrafast metabolizer phenotype showing higher-than-average enzymatic activity [3–6].

The clinical importance of genotyping is most obvious in drug therapies with a high and poorly predicted rate of non-responders, as well as in drug therapies with narrow therapeutic indices, where severe side effects can occur. For example, in depression 30% of patients do not respond significantly to psychopharmacological therapy, and the clinical efficacy of therapy can usually only be evaluated after at least 2 weeks of drug administration. Here, additional factors that help to predict individual drug response are of potential clinical significance (such as the actual metabolic activity measured by therapeutic drug monitoring, previous treatment success, and the influence of genetic factors). Genetically caused differences in drug-metabolizing enzyme activity may be one important point to consider when striving to optimize treatment strategies. However, the categorization of clinical phenotypes by genotype still does not account for the wide variations in individual patient risk.

Considering Individual Drug Metabolism: Rationale behind the Development of *CYP2D6* Testing

Genetically caused variability in drug metabolism is reflected in differences in clearance, half-life and trough plasma concentrations. These pharmacokinetic differences are highly replicable and can be used for genotype-based dose adjustments if a direct correlation with clinical outcomes is demonstrated. Methods for extracting dose adjustments from genotypes have been developed and published elsewhere [7–12].

Differences in pharmacokinetic parameters such as oral clearances may be overcome by adjustment of the drug dose. These dose adjustments make sense for those drug therapies where there is a predictable association between plasma concentration and clinical efficacy (such as antibiotics and proton pump inhibitors). However, for some drugs (such as antidepressants), recorded plasma concentrations correlate poorly with clinical efficacy.

Where the systemic concentration or target tissue exposure in psychiatric drug treatment is not directly accessible, the drug plasma levels might not correlate well with target tissue exposure, with the consequence that pharmacokinetics is thus not the 'rate-limiting step' in achieving drug efficacy. The correlation between plasma concentration and clinical efficacy depends on how the pharmacological action of the drug is influenced by the kinetics of the interaction between drug and receptor and on other downstream effects or disease mechanisms.

In addition, clinical assessment instruments in psychiatry are rather crude and cast little light on the underlying etiology. It is therefore important to note that difficulties in the prediction of individual drug response originate not only in our lack of knowledge of individual patient factors, but also in the nature of the clinical endpoints used in research and therapeutics [13, 14].

As a consequence, empirical dose-finding strategies based on clinical observation may be used to obtain optimal response and to avoid adverse drug effects.

In figure 1, the differences in mean oral clearance between carriers of none, one, two and more active *CYP2D6* genes are expressed as percent of dose adjustments for

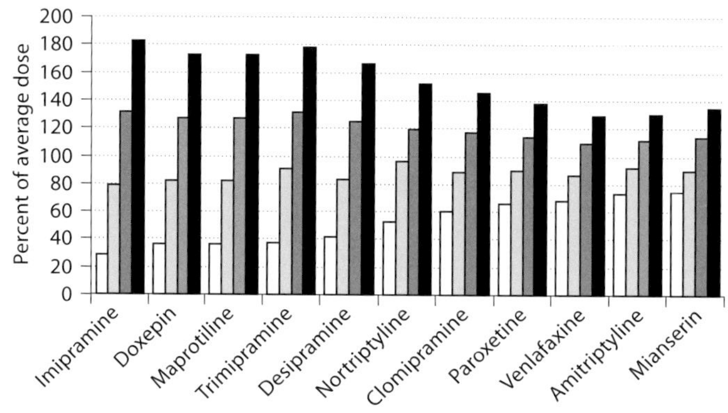

Fig. 1. *CYP2D6*-genotype-dependent quantitative changes in pharmacokinetics of antidepressant drugs expressed as percent dose adaptations. CYP2D6 poor metabolizer (white), intermediate metabolizer (light gray), extensive metabolizer (dark gray) and ultrafast metabolizer (black). Reprinted with permission from Kirchheiner and Seeringer [44]. Dose adaptations were calculated based on published pharmacokinetic data in dependence on the *CYP2D6* genotype as described in Kirchheiner et al. [45]. Dose adaptations are based on an average dose of 100% and are targeted at the Caucasian population. Data from studies in Asian or African or other populations were not incorporated, since poor metabolizer data were lacking.

antidepressants. In this figure, a significant variation is seen between the extent to which different substrates are cleared by the *CYP2D6* copy number variants. Hence, the decision regarding whether a genotype-based dose adjustment might be clinically important or not is dependent on the extent to which CYP2D6 metabolizes a particular substrate. The involvement of multiple elimination routes in the metabolism of a particular substrate and substrate interaction with other metabolic enzymes will decrease the importance of a single-gene polymorphism. However, if the systemic concentration of the parent drug or metabolite is critical for reasons of drug safety, polymorphic drug metabolisms might alter the individual risk of side effects. In this case, genotyping might suggest using an alternative drug. This approach has proven useful for drugs like venlafaxine and risperidone where CYP2D6-mediated metabolism produces equally potent metabolites. In this example, the risk of side effects in poor CYP2D6 metabolizers is higher than that of fast CYP2D6 metabolizers even when the sum of the parent drug and metabolite concentrations was the same [15–18].

Limitations of Pharmacogenetic Dose Adjustments

When calculating pharmacokinetic-dependent dose adjustments, several important issues must be considered. Often, metabolites are produced that are also pharmacologically active, and this might influence therapeutic efficacy or the chances of adverse effects. Hence, the existence of significant circulating concentrations of metabolites should be taken into account in the dosing adjustments. This may be achieved either by adding drug and metabolite(s) to get a total active moiety, or by recommending a change in drug choice (if metabolites are generated which have potential for adverse drug effects). The fate of these metabolites generally depends on other (sometimes also polymorphic) enzymes, and therefore the individual risk of side effects might also vary.

Of the existing studies available that have considered drug-metabolite(s) augmentation, only those based on large samples have the precision required to provide recommendations for dose adjustments. However, in the majority of cases, information is available only from small-size trials in healthy volunteers, and is derived from only a handful of homozygous carriers of a particular drug-metabolizing genetic variant. Furthermore, it is important that the dose range used in these kinds of studies should be set in the clinical range. Often, this is not the case, with many studies using healthy volunteers and low drug doses. However, dose recommendations cannot be extrapolated automatically to the dose range used in patients.

A large sample size is required to estimate whether the influence of a genetic variant on drug metabolism and benefit/risk ratio is significant, or whether it is just one of a number of additional factors that contribute to drug efficacy. For example, comedication has to be taken into account, since genotypic variation can also lead to differences in target drug efficacy. Of course, the enzyme activity of a genetic poor metabolizer cannot be effectively increased or decreased by substances which are inducers or inhibitors of the enzyme as it is absent. However, extensive or ultrarapid metabolizers can be 'converted' to the poor metabolizer phenotype by strong enzyme inhibitors. Other pharmacogenetic variants may also influence the effects of the *CYP2D6* variants. Drug transport [19] or genetic variability in drug targets might confound the effects caused by *CYP2D6* variants and modify the dose selection in the individual patient.

In order to reduce the 'noise' in any study seeking to elucidate the relationship between plasma concentration and impact on benefit/risk, pharmacological interactions that involve the drug in question and known inhibitors which indirectly alter drug efficacy might be useful. However, one problem in this respect is that inhibitors are often unspecific and affect more than one metabolic pathway. Therefore, interactions between known inhibitors and drug-metabolizing enzymes might not completely represent the pharmacokinetics/dynamics of a genetically poor metabolizer.

Prospective validation of genotype-based dose adjustments is necessary, and several studies are now being performed that compare therapy with pharmacogenetic diagnostics with standard therapy in a randomized controlled fashion. These prospective trials and other trials will only be useful in routine clinical practice if they demonstrate the validity, utility and cost-effectiveness of pharmacogenetic testing [20].

Pharmacogenetic Study Designs Validating Predictive Markers for Drug Response

Randomized controlled studies provide the most stringent proof-of-concept for pharmacogenetically based optimization of drug therapy. Such studies on pharmacogenetic diagnostics are rather expensive and usually involve some form of multicenter design. A controlled pharmacogenetic prospective trial is usually suitable for evaluating common drug therapies and where there are rather frequent genetic variants. At the moment, several international multicenter trials on the prospective evaluation of genotype-specific dosing are in progress in several fields of drug therapy, for example, oral anticoagulants, tamoxifen, abacavir, isoniazid, proton pump inhibitors and azathioprine. In the field of psychotropic drug treatment, no randomized prospective pharmacogenomic study on genotype-specific dosing has yet been published. This is probably due to two reasons: the lack of correlation between dose and drug response (a problem affecting antidepressant drug therapy), and the feasibility of a clinical study requiring a large sample size of patients in monotherapy.

Classification and Integration of Biomarkers as Surrogate Endpoints in Clinical Efficacy Trials

As noted, the importance of genotypic variation in psychiatric practice may be limited by the lack of a clear association between plasma drug levels and clinical efficacy. This lack of association may be due to the existence of a large number of intermediate factors affecting the efficacy of treatment. If this is the case, a strategy based on designing clinical studies monitoring treatment outcome may not be advisable. If the underlying mechanisms are complex enough, the detection of the multiple factors affecting a qualitative outcome (such as response or emergence of adverse effects) cannot be achieved by observing statistical correlations alone.

An alternative strategy would aim at elucidating the biochemical processes generated by antecedent genetic polymorphisms, especially those that give rise to the outcome measures (fig. 2). The rationale for this strategy is based on the assumption that these intermediaries may correlate with the clinical endpoint, and are representative of the net effect of the genetic polymorphism on the clinical efficacy endpoint. Spe-

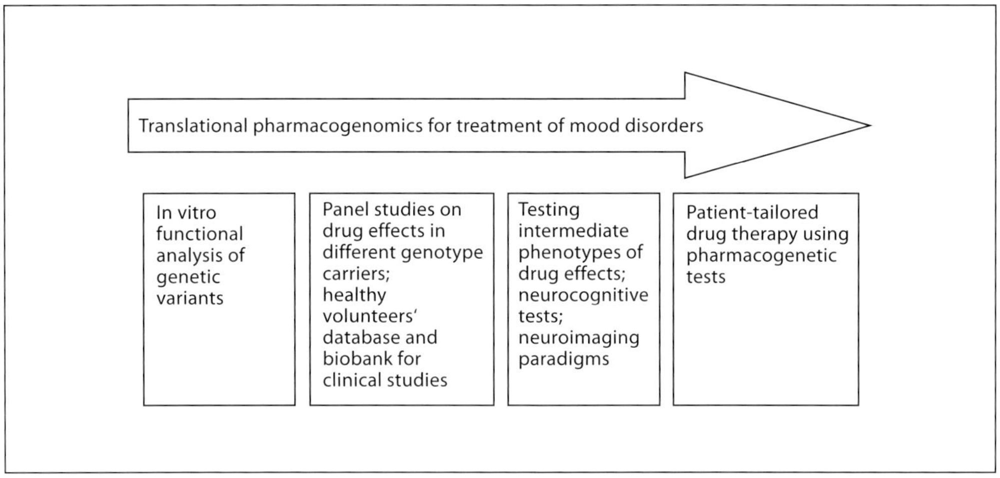

Fig. 2. Progression of studies, from basic to applied, in translational pharmacogenomics. Consideration of the most valid rational study designs for different situations in drug therapy might lead to a more effective translation of pharmacogenetics into clinical practice.

cific surrogate markers for drug response which are related to the function of a gene of interest have been called intermediate phenotypes or endophenotypes [21]. The use of endophenotypes as an outcome parameter in clinical pharmacogenomic studies directly serves the aim of individualizing drug therapy.

Numerous potential biomarkers or intermediate phenotypes exist that could influence a diffuse clinical phenotype, such as depression, and there is a need to qualify a potential biomarker for use in clinical practice. Biomarker qualification involves the collation of evidence linking a biomarker with disease biology and the clinical endpoint. The qualification process includes several general phases, leading from exploration to demonstrations, further characterization and, finally, to the evaluation of the substitute character of a biomarker for a clinical endpoint [22].

One example demonstrating the potential of intermediate phenotypes for understanding the relation between disorder subtypes, individual genetic makeup and psychopharmacological interventions is given by depression [23]. Although not the sole abnormality found in depression, alterations in monoaminergic function are likely to play a key role in the pathogenesis of this disorder, as demonstrated by the reliability and specificity of the induction of depressive symptoms by tryptophan depletion [24, 25]. Current interest has focused on studies linking monoaminergic function with areas of the brain specifically involved in the processing of emotions and dispositional mood. Examples in this respect are the roles that dopaminergic, serotonergic and noradrenergic systems play in processing reward salience [26], adaptability (in conditions such as reversal learning [27]) and memory for emotionally salient events [28], respectively. Neuroimaging combined with genetics represents a potentially powerful approach for closing the genotype-endophenotype-phenotype circle [29].

Neuroimaging studies of depression have already identified a number of brain regions showing differential changes in patients versus controls when at rest and when processing emotional stimuli [30, 31], and studies of familial and melancholic subtypes of depression have found rest perfusion and metabolism in the amygdale to be elevated in patients versus controls [32–34], a part of the limbic system that processes emotionally salient information [35]. Furthermore, the amygdala was found to react more strongly to emotional stimuli in depressed individuals [36, 37]. Neuroimaging studies provide some evidence that the amygdala and the ventromedial prefrontal cortex might have a role in the neurobiology of depression. These two brain structures are thought to constitute feedback circuits in which the reactivity to emotionally relevant stimuli is modulated by control signals elaborated in the prefrontal cortex, and which may be altered in depression [38, 39].

However, it has to be said that the data emerging from neuroimaging studies of endophenotypes in depression is not consistent. The absence of control groups, lack of randomization, failure to adjust data to take account of differing response rates, use of uncorrected significance levels and the absence of some valid assessment instrument for inferring differences between groups have all added to the incoherence of the neuroimaging data.

Drug target exposure is central to the assessment of clinical drug-related response. To date, most of the available clinical studies have not included such an assessment. Neuroimaging techniques, in particular PET, can provide evidence for the relationship between target exposure, response and genotype. However, PET is not a suitable instrument for monitoring course of illness or clinical response in the longer term, and recent studies have instead used functional MRI to correlate changes in areas associated with depression to candidate (functional) genetic polymorphisms [40, 41].

Data is starting to emerge from neuroimaging-molecular genetics studies of depression which suggests that such approaches have a significant role to play in understanding the etiology of this complex disorder. However, as with all studies of associations that combine semi-qualitative outcome variables with quantitative genotypes, careful study design is needed if these studies are to be used to underpin better therapeutic interventions.

More work is needed to develop perfusion techniques with sufficient precision [42] and to develop neuroimaging statistical techniques that quantify deviations from normative values [43]. Beyond the use of endophenotypic biological markers, longitudinal data such as hospitalization time (episodal course), quality of life, disability or survival are warranted in order to finally estimate the cost/benefit ratio of pharmacogenetic testing.

Conclusion

Currently, pharmacogenomic testing in psychiatry is characterized by a lack of reliable biomarkers predicting individual drug reactions. As a result, prescribing is often based on a 'trial and error' approach. Whilst the poor predictive value of genes might be seen as a consequence of the lack of understanding of how different etiologies and organic changes contribute to the complex syndrome depression, new approaches that combine genetics with endophenotypes show some promise.

Currently, genotyping psychiatric patients for *CYP2D6* variants represents the only example of a pharmacogenetic test widely accepted and more or less used in clinical psychiatric practice. However, this genotype only allows the prediction of drug metabolism kinetics. Genetic predictors of drug response have been searched for intensively, but, to date, none have shown any predictive value. Accurate predictions of treatment response in multifactorial diseases are likely to require an understanding of the contributions made by a number of intermediate phenotypes towards the etiology of the disease state. Functional neuroimaging studies show promise in the detection of endophenotypic markers in psychiatric disorders such as depression, but are not at the stage where they can be applied in clinical practice.

To achieve the translation of pharmacogenetic data into clinical practice, a variety of clinical studies with different study designs are required. Integrative approaches to data analysis are necessary to obtain the broad critical mass of evidence needed to convince clinicians and healthcare providers to use and pay for pharmacogenetic diagnostics as a tool to improve and individualize patient treatment in psychiatry.

References

1 Evans WE, Relling MV: Pharmacogenomics: translating functional genomics into rational therapeutics. Science 1999;286:487–491.

2 Bertilsson L, Dahl ML, Dalen P, Al-Shurbaji A: Molecular genetics of CYP2D6: clinical relevance with focus on psychotropic drugs. Br J Clin Pharmacol 2002;53:111–122.

3 Sachse C, Brockmöller J, Bauer S, Roots I: Cytochrome P450 2D6 variants in a Caucasian population: allele frequencies and phenotypic consequences. Am J Hum Genet 1997;60:284–295.

4 Zanger UM, Fischer J, Raimundo S, Stuven T, Evert BO, Schwab M, et al: Comprehensive analysis of the genetic factors determining expression and function of hepatic CYP2D6. Pharmacogenetics 2001; 11:573–585.

5 Griese EU, Zanger UM, Brudermanns U, Gaedigk A, Mikus G, Morike K, et al: Assessment of the predictive power of genotypes for the in-vivo catalytic function of CYP2D6 in a German population. Pharmacogenetics 1998;8:15–26.

6 Gaedigk A, Ndjountche L, Divakaran K, Dianne Bradford L, Zineh I, Oberlander TF, et al: Cytochrome P4502D6 (CYP2D6) gene locus heterogeneity: characterization of gene duplication events. Clin Pharmacol Ther 2007;81:242–251.

7 Kirchheiner J, Fuhr U, Brockmöller J: Pharmacogenetics-based therapeutic recommendations – ready for clinical practice? Nat Rev Drug Discov 2005;4: 639–647.

8 Kirchheiner J, Brosen K, Dahl ML, Gram LF, Kasper S, Roots I, et al: CYP2D6 and CYP2C19 genotype-based dose recommendations for antidepressants: a first step towards subpopulation-specific dosages. Acta Psychiatr Scand 2001;104:173–192.

9 McLeod HL, Siva C: The thiopurine S-methyltransferase gene locus – implications for clinical pharmacogenomics. Pharmacogenomics 2002;3:89–98.
10 Wu AH, Wang P, Smith A, Haller C, Drake K, Linder M, et al: Dosing algorithm for warfarin using CYP2C9 and VKORC1 genotyping from a multi-ethnic population: comparison with other equations. Pharmacogenomics 2008;9:169–178.
11 Wen MS, Lee M, Chen JJ, Chuang HP, Lu LS, Chen CH, et al: Prospective study of warfarin dosage requirements based on CYP2C9 and VKORC1 genotypes. Clin Pharmacol Ther 2008;84:83–89.
12 Sconce EA, Khan TI, Wynne HA, Avery P, Monkhouse L, King BP, et al: The impact of CYP2C9 and VKORC1 genetic polymorphism and patient characteristics upon warfarin dose requirements: proposal for a new dosing regimen. Blood 2005;106: 2329–2333.
13 Katz MM, Tekell JL, Bowden CL, Brannan S, Houston JP, Berman N, et al: Onset and early behavioral effects of pharmacologically different antidepressants and placebo in depression. Neuropsychopharmacology 2004;29:566–579.
14 Trivedi MH, Morris DW, Grannemann BD, Mahadi S: Symptom clusters as predictors of late response to antidepressant treatment. J Clin Psychiatry 2005;66:1064–1070.
15 Troost PW, Lahuis BE, Hermans MH, Buitelaar JK, van Engeland H, Scahill L, et al: Prolactin release in children treated with risperidone: impact and role of CYP2D6 metabolism. J Clin Psychopharmacol 2007;27:52–57.
16 Bork JA, Rogers T, Wedlund PJ, de Leon J: A pilot study on risperidone metabolism: the role of cytochromes P450 2D6 and 3A. J Clin Psychiatry 1999; 60:469–476.
17 Shams ME, Arneth B, Hiemke C, Dragicevic A, Muller MJ, Kaiser R, et al: CYP2D6 polymorphism and clinical effect of the antidepressant venlafaxine. J Clin Pharm Ther 2006;31:493–502.
18 Lessard E, Yessine M, Hamelin B, O'Hara G, LeBlanc J, Turgeon J: Influence of CYP2D6 activity on the disposition and cardiovascular toxicity of the antidepressant agent venlafaxine in humans. Pharmacogenetics 1999;9:435–443.
19 Uhr M, Tontsch A, Namendorf C, Ripke S, Lucae S, Ising M, et al: Polymorphisms in the drug transporter gene ABCB1 predict antidepressant treatment response in depression. Neuron 2008;57:203–209.
20 Grossman I: Routine pharmacogenetic testing in clinical practice: dream or reality? Pharmacogenomics 2007;8:1449–1459.
21 Gottesman II, Gould TD: The endophenotype concept in psychiatry: etymology and strategic intentions. Am J Psychiatry 2003;160:636–645.
22 Wagner JA, Williams SA, Webster CJ: Biomarkers and surrogate end points for fit-for-purpose development and regulatory evaluation of new drugs. Clin Pharmacol Ther 2007;81:104–107.
23 Hasler G, Drevets WC, Manji HK, Charney DS: Discovering endophenotypes for major depression. Neuropsychopharmacology 2004;29:1765–1781.
24 Bell C, Abrams J, Nutt D: Tryptophan depletion and its implications for psychiatry. Br J Psychiatry 2001;178:399–405.
25 Rogers RD, Tunbridge EM, Bhagwagar Z, Drevets WC, Sahakian BJ, Carter CS: Tryptophan depletion alters the decision-making of healthy volunteers through altered processing of reward cues. Neuropsychopharmacology 2003;28:153–162.
26 Cools R: Role of dopamine in the motivational and cognitive control of behavior. Neuroscientist 2008; 14:381–395.
27 Robbins TW: Synthesizing schizophrenia: a bottom-up, symptomatic approach. Schizophr Bull 2005;31:854–864.
28 Strange BA, Hurlemann R, Dolan RJ: An emotion-induced retrograde amnesia in humans is amygdala- and beta-adrenergic-dependent. Proc Natl Acad Sci USA 2003;100:13626–13631.
29 Meyer-Lindenberg A, Weinberger DR: Intermediate phenotypes and genetic mechanisms of psychiatric disorders. Nat Rev Neurosci 2006;7:818–827.
30 Drevets WC: Neuroimaging studies of mood disorders. Biol Psychiatry 2000;48:813–829.
31 Mayberg HS: Positron emission tomography imaging in depression: a neural systems perspective. Neuroimaging Clin N Am 2003;13:805–815.
32 Drevets WC, Videen TO, Price JL, Preskorn SH, Carmichael ST, Raichle ME: A functional anatomical study of unipolar depression. J Neurosci 1992; 12:3628–3641.
33 Whalen PJ, Shin LM, Somerville LH, McLean AA, Kim H: Functional neuroimaging studies of the amygdala in depression. Semin Clin Neuropsychiatry 2002;7:234–242.
34 Drevets WC, Price JL, Bardgett ME, Reich T, Todd RD, Raichle ME: Glucose metabolism in the amygdala in depression: relationship to diagnostic subtype and plasma cortisol levels. Pharmacol Biochem Behav 2002;71:431–447.
35 Morris JS, Frith CD, Perrett DI, Rowland D, Young AW, Calder AJ, et al: A differential neural response in the human amygdala to fearful and happy facial expressions. Nature 1996;383:812–815.
36 Sheline YI, Barch DM, Donnelly JM, Ollinger JM, Snyder AZ, Mintun MA: Increased amygdala response to masked emotional faces in depressed subjects resolves with antidepressant treatment: an fMRI study. Biol Psychiatry 2001;50: 651–658.

37 Siegle GJ, Steinhauer SR, Thase ME, Stenger VA, Carter CS: Can't shake that feeling: event-related fMRI assessment of sustained amygdala activity in response to emotional information in depressed individuals. Biol Psychiatry 2002;51:693–707.
38 Sotres-Bayon F, Bush DE, LeDoux JE: Emotional perseveration: an update on prefrontal-amygdala interactions in fear extinction. Learn Mem 2004;11: 525–535.
39 Phillips ML, Drevets WC, Rauch SL, Lane R: Neurobiology of emotion perception. I. The neural basis of normal emotion perception. Biol Psychiatry 2003;54:504–514.
40 Lesch KP, Mossner R: Genetically driven variation in serotonin uptake: is there a link to affective spectrum, neurodevelopmental, and neurodegenerative disorders? Biol Psychiatry 1998;44:179–192.
41 Caspi A, Sugden K, Moffitt TE, Taylor A, Craig IW, Harrington H, et al: Influence of life stress on depression: moderation by a polymorphism in the 5-HTT gene. Science 2003;301:386–389.
42 Viviani R, Sim EJ, Lo H, Richter S, Haffer S, Osterfeld N, Thöne J, Beschoner P: Components of variance in brain perfusion and the design of studies of individual differences: the baseline study. Neuroimage 2009;46:12–22.
43 Viviani R, Beschoner P, Jaeckle T, Hipp P, Kassubek J, Schmitz B: The bootstrap and cross-validation in neuroimaging applications: estimation of the distribution of extrema of random fields for single volume tests, with an application to ADC maps. Hum Brain Mapp 2007;28:1075–1088.
44 Kirchheiner J, Seeringer A: Clinical implications of pharmacogenetics of cytochrome P450 drug metabolizing enzymes. Biochim Biophys Acta 2007; 1770:489–494.
45 Kirchheiner J, Nickchen K, Bauer M, Wong ML, Licinio J, Roots I, et al: Pharmacogenetics of antidepressants and antipsychotics: the contribution of allelic variations to the phenotype of drug response. Mol Psychiatry 2004;9:442–473.

Prof. Julia Kirchheiner
Institute of Pharmacology of Natural Products and Clinical Pharmacology, University of Ulm
Helmholtzstrasse 20
DE–89081 Ulm (Germany)
Tel. +49 731 500 65603, Fax +49 731 500 65605, E-Mail julia.kirchheiner@uni-ulm.de

Schwab M, Kaschka WP, Spina E (eds): Pharmacogenomics in Psychiatry.
Adv Biol Psychiatry. Basel, Karger, 2010, vol 25, pp 12–29

Pharmacogenomics: Reflecting on the Old and New Social, Ethical and Policy Issues in Postgenomics Medicine

Vural Ozdemir

Department of Social and Preventive Medicine, School of Public Health, University of Montreal, Montreal, Que., Canada

Abstract

Whenever a new form of biotechnology is introduced, there is often uncertainty around its intended and unintended impacts on science, medicine and society. Past experiences with genetically modified organisms, stem cell research and other health technologies have taught us some important lessons – that it is not just scientific and technical factors that are important to the uptake of innovative technologies. In the case of pharmacogenomics, a field of inquiry that aims to discern the genomic basis of individual and population differences in drug effects, there has been much written on the attendant promises and limitations. Since the completion of the Human Genome Project in 2003, we are, however, in the postgenomics era. This brings some of the 'old questions' that remained unaddressed in pharmacogenomics to the forefront, e.g. race-based therapeutics. Moreover, 'new questions' in postgenomics medicine – such as privacy and confidentiality in hypothesis-free genome-wide association studies, and regulation of direct-to-consumer personal genomics tests – require critical reexamination of the established practices in both biosciences and bioethics. While other chapters in this book aim to address the technical and scientific factors, the present chapter presents an analyses of the old and new social, ethical and policy issues that can impact the uptake of pharmacogenomic innovations and their future trajectory in postgenomics medicine.

Historical Context and the Knowledge Gaps to Be Addressed

Pharmacogenomics has its origins in the 1950s with the early observations on monogenic variations in drug metabolism. This long history is closely intertwined with psychiatry. First, a technical factor that catalyzed the advances in both pharmacogenomics and psychiatric drug therapy was the introduction and availability of modern high-performance chromatography methods (e.g. HPLC), particularly in the 1970s. Arguably, HPLC was then akin to the high-throughput genome sequencers of the present day pharmacogenomics laboratories. These technical breakthroughs allowed the measurement of drug and endogenous metabolite concentrations in biological fluids, e.g. in the plasma and the cerebrospinal fluid, and thus for the first time offered insights into the intermediary steps between psychiatric drug administration and the appearance of a therapeutic or toxic drug effect [1–3]. Prior to adoption of such laboratory methods, mechanistic steps in drug disposition essentially remained as a 'black box' to most clinical psychiatrists. These quantitative pharmacokinetic measures also provided the much needed pharmacological phenotypes for association studies using human genetic variation data. This led to recognition of the fact that the dispositions of some drugs display polymorphic distributions, raising the possible influence of genetic factors. Many of the subsequent studies in pharmacogenetics dealt with the hereditary basis of person-to-person differences in psychiatric drug disposition, response and toxicity [4]. Second, the 'biological turn' in psychiatry – instead of psychoanalytical approaches – in the second half of the 20th century and the launch of the Human Genome Project (HGP) in the early 1990s coincided with a shared emphasis on biological determinants of drug effects. These events collectively contributed to convergence of the two fields as psychiatric pharmacogenetics over the past decade.

In addition to these shifts in research priorities and technical advances over the past five decades, both psychiatry and pharmacogenomics have been further shaped by a growing interest and concerns for the ethical, legal and social issues (ELSIs) emerging from the HGP [5]. The US Department of Energy and the National Institutes of Health allocated 3–5% of their annual HGP budgets toward studying the ELSIs surrounding the availability of genetic information. Enabled in part by this marked funding support, the ELSI program has rapidly gained momentum since the 1990s. The US ELSI program had a number of priority areas, such as fairness in the use of genetic information (e.g. by insurers and employers), privacy and confidentiality of genetic information, societal impacts and stigmatization due to human genetic differences, adequate informed consent, clinical genetic testing and commercialization of products [5, 6]. In Europe, a similar transdisciplinary program on the ethical, legal and social aspects (ELSA) was established by the European Commission in 1994 [7]. Unlike the US ELSI framework, the European ELSA program aimed to address not only genetics/genomics but the broader field of life sciences and technologies [7].

The professional scope of bioethics as an integral component of the HGP, whether in the form of ELSI or ELSA activities, began to increasingly overlap with and extend into science governance, public policy, health technology assessment, and the study of innovation processes in the larger context of knowledge-based national economies. These newer roles and responsibilities assigned to bioethics – in addition to the previous traditional focus on protection of research subjects and rules governing the doctor and patient relationships – make it obvious that bioethics is a subject that cannot be neglected in discussions of new technologies and scientific innovations. As noted later in this chapter, this also calls, however, for greater empiricism and provision of evidence regarding the claims made in and by bioethics (i.e. evidence-based ethics) and importantly, study of the social, political, economic, scientific and technical contexts and the 'fault lines' that shape and are shaped by the emerging bioethics issues [8–12]. Without a firm empirical grasp of the latter contexts in which bioethics is 'embedded', it is difficult to conceive, develop and test evidence-based policy options to address the real-life bioethics issues actually impacting the individuals and populations.

When the ideals or norms prescribed by the traditional philosophical bioethics principles (see the following section) do not coincide with the actual realities of scientific practice, there is a certain degree of responsibility to work towards effective solutions and interventions, e.g. in the form of public policies or regulatory oversight, in order to remedy disconnects between ethical standards and scientific or clinical practice. Hitherto, this has not materialized appreciably in part because bioethics analyses tended to lack an empirical study of the lived contexts associated with the ethical issues; furthermore, bioscientists have not pursued empirical bioethics research in large enough numbers to form a critical mass to allow the discernment of scientific and technical nuances that have bioethics significance. These dilemmas, resource limitations and methodological conflicts within the bioethics field itself remain largely opaque to the pharmacogenomics and genomics research community, a knowledge gap that the present chapter aims to address. The personal view of this author is that scientists tend to perceive bioethicists as being engaged in primarily philosophical inquiries, with incomplete recognition of the marked heterogeneity in bioethics decision-making processes and the attendant need for an empirical and evidentiary basis to bioethics. This idea is in part supported by recent qualitative research that posits that genomics scientists, while considering bioethics as an important function, tend to delegate this role to the bioethicists rather than extending the scope of their self-governance in scientific practice to bioethical reasoning as well [13].

As we move towards an era of genome-wide association studies that demand hypothesis-free research and informed consent for future unspecified use of biological samples while ensuring research subjects' protection in psychiatry and the proliferation of public and private stakeholders in personalized medicine with divergent and often conflicting interests that are not always immediately obvious, can we afford to be passive and simply extend the previous practices of bioethics to the postgenomics era? Alternatively, has the time come for a critical reexamination of both pharmacogenomics science and bioethics to bring about further transparency and better integration of these knowledge domains in psychiatry?

Keeping in mind these overarching challenges from within the practice of pharmacogenomics science and bioethics, and the tension zones at their intersections, the present chapter has 3 primary aims:

(1) To introduce the reader to the principle-based approach that dominated bioethical reasoning in clinical research and practice, particularly in North America, and explain why the context in which a bioethical issue is embedded also matters.
(2) To present a synopsis of the old and new social, ethical and policy issues associated with the practice of pharmacogenomics science in postgenomics medicine.
(3) To examine a long-standing unresolved socio-ethical and policy controversy, i.e. race-based pharmacogenomics. Because race has often been used as a stratification axis in clinical psychopharmacology and psychiatric pharmacogenomics, this is a crucial issue that needs in-depth evaluation in the postgenomics era.

A clear understanding of the above dimensions is essential for the readers to independently and critically evaluate the socio-ethical and policy issues as they emerge in the practice of psychiatric pharmacogenomics in the postgenomics era. Importantly, throughout the chapter, we present a future outlook and the 'next-generation questions' within bioethics and, in particular, considerations for 'ethics of bioethics' by recognizing the history, processes and values associated with the choice of philosophical principles used in bioethics decisions, the role of bioscientists and social scientists in identifying the regional and global contexts – whether they be scientific, technical, social, economic or political – that can decisively impact the future practice of evidence-based bioethics in psychiatric pharmacogenomics.

20th Century Biomedical Ethics and the Principle-Based Approach to Decision-Making in Bioethics

Before one can critically examine the ethical implications of pharmacogenomics raised by several recent transdisciplinary committee reports [14–17], we need to recall the brief history and background on the dominant decision-making processes used in biomedical ethics in the 20th century. This is essential in order to discern the strengths and shortcomings of the previous bioethics analyses on an increasingly globalized pharmacogenomics research.

Biomedical ethics and research ethics rose to prominence in the second half of the 20th century as a response to crimes against humanity and the abuse of human subjects in medical research [18, 19]. In the Tuskegee syphilis study (1932–1972), 600 African-American men were enrolled to investigate the racial differences in the natural course of untreated syphilis. This ethical breach culminated in the Belmont Report issued by the US National Commission for the Protection of Human Subjects of Biomedical and Behavioral Research [19]. Since then, decision-making in the field of biomedical and research ethics has primarily relied on the theme of 'protection' and normative analyses using moral principles such as autonomy, beneficence, nonmaleficence and justice [20]. This approach is also known as the 'four principles approach' and has its basis in the Belmont Report [19] and the works of Childress and

Beauchamp first presented in their textbook *Principles of Medical Bioethics* in 1979 [20]. These four principles have been frequently used as guidance for setting the bioethics norms in both research and clinical practice:

- respect for autonomy: respecting the decision-making capacities of autonomous persons;
- beneficence: balancing benefits against the risks and costs;
- nonmaleficence: avoiding the causation of harm;
- justice: distributing benefits, risks and costs fairly.

As it might be already clear to the reader, bioethics analyses have tended to concentrate on norms and standards concerning how things (e.g. a certain scientific or clinical practice) 'ought to be' drawing from moral theories and philosophical principles, but did not always adequately recognize how 'things actually are' in practice, or the social, political, economic, scientific and technical contexts in which the bioethics dilemmas emerge. This asymmetry in bioethics, i.e. greater recognition of the idealized norms over their practice and the attendant social context in a real-life setting, however, isolates bioethics from the actual practice of science and medicine [8, 9, 11, 12, 21, 22]. The intent of this critique is not to say that the above four principles are not useful – they are – but neglecting the real-life context of science and medicine results in a prescriptive and top-down hegemony in bioethical reasoning, and eliminates the possibility of designing effective policy interventions to bring together the lived practice of science and medicine with the bioethics norms when the practice and theory are not aligned (and often they are not). It is therefore not surprising that the words 'evidence and data' still remain as exotic and unfamiliar terms within the field of bioethics, but they need consideration if evidence-based bioethics will make a contribution to bridge the gaps between the theory and the practice of bioethics.

As we note in the next section of this chapter, there is little population-based empirical data on the actual range and impact of the social, ethical and policy issues within the field of pharmacogenomics. This author therefore contends that there is a need for a new subfield of pharmacogenomics dealing with 'ethics epidemiology'. This is not a far-fetched idea as already the rationale for 'pharmacogenomics epidemiology' has been presented elsewhere [23]. The latter new hybrid field assesses the range of responses to pharmacologic agents in relation to genetic variation in population groups [23, 24].

In contemporary bioethics analyses, one often reads statements classifying a behavior or practice as ethical or unethical, but the process for this decision-making is not always clearly stated and nor do the bioethics principles utilized to draw such conclusions operate in a vacuum. That is, bioethics principles may often compete and conflict with each other (e.g. consider the principles of 'autonomy' and 'solidarity'). In western bioethics practice, autonomy has come to prevail over-and-above other bioethics principles, particularly in the USA [22, 25]. This can be problematic as pharmacogenomics research is now truly global in nature, crossing national, political, geographical and cultural borders. Many countries in the Asia-Pacific region are becoming powerful contributors to genomics and personalized medicine science. While autonomy may be appealing when viewed from a western context, other principles may be more important in different societies and contexts. This is not to advocate for moral relativism, but instead to merely express that the context matters in a bioethics analysis. Moreover, different principles may be necessary in research concerning different communities. For example, in the process of collection of genetic samples from the South Pacific island of Tonga, the individual informed consent procedure was met with opposition for failing to address the traditional Tongan role of the extended family in decision-making [26]. As we look further into the future, more complex scenarios will likely emerge when conducting research with immigrant populations living and working in diaspora who often have to sustain their existence in cultures and value systems vastly different than their native countries. The lesson for our purposes is that even though the concept of 'harmonization' is intuitively appealing from a laboratory or technical process standpoint, there will be local and regional ethical principles operational in different communities in the face of an increasingly globalized genomics research. Hence, caution and in-depth considerations are necessary when using the term 'harmonization' in reference to the bioethics principles, in contrast to well-justified technical harmonization necessary in scientific experiments.

In pharmacokinetic calculations, it is customary to expressly indicate which pharmacokinetic model is employed in derivation of the system parameters, such as drug clearance or volume of distribution. In the case of traditional philosophical bioethics reasoning, there is also a moral significance attached to which particular bioethics principles are chosen over others, or how different principles are weighed and compared to each other when they compete or conflict. Even though on first glance the world of pharmacokinetic modeling and bioethics reasoning may appear entirely unrelated, they indeed share the need for transparency and accountability in being explicit about such 'upstream factors' in decision-making process, i.e. the choice and weighing of different normative bioethics principles instead of stating merely that a scientific practice or health technology is ethical or unethical.

Finally, it is worth emphasizing that the philosophical bioethics that prevailed thus far can only benefit from engagement with social scientists and bioscientists (and vice versa). Theoretical and reasoned discussions in bioethics are also important, but they need to be grounded in the nuances of scientific and medical practice as noted above. This can help to achieve a more contextualized and lived moral reasoning in bioethics, whether it concerns pharmacogenomics or other new omics technologies that will soon appear on the horizon.

Old and New Social, Ethical and Policy Issues at the Junction of Pharmacogenomics and Postgenomics Medicine

Past experience with genetically modified organisms, stem cell research and other health technologies have taught us some important lessons – that it is not just scientific and technical factors that are important to the uptake of innovative technologies. There has been recognition early on that the course of development and implementation of pharmacogenomics will also depend on identifying and addressing the social, ethical and policy issues. Several reports aimed to address such factors that are crucial for a productive and just direction in pharmacogenomics or genomic medicine [14–17, 27, 28]. We herein focus on the report by the Nuffield Council on Bioethics [15], which was among the first to attempt to define the ethical, social and policy issues that pharmacogenomics might raise; these are applicable to both developed and developing countries [29]. We also discuss some of the newer relevant issues in postgenomics medicine and refer the reader to other pertinent literature.

The Nuffield Council on Bioethics usefully framed the social, ethical and policy issues on pharmacogenomics around 4 categories:

Category 1: Information
This is also the category that attracted considerable attention in bioethics literature, including concerns over informed consent, privacy, confidentiality, anonymity, disclosure of genetic information and potential for discrimination. The nature of pharmacogenomics information has been first evaluated with respect to the concept of 'genetic exceptionalism': the view that genetic tests are categorically distinct from medical tests that do not concern DNA, and that they therefore raise different ethical issues. The Nuffield Council on Bioethics recognizes that both genetic and non-genetic tests can be rich in information, and thus 'there is no reason to assume that genetic information, including pharmacogenetic information, is qualitatively different from other medical information' [15].

Category 2: Resources
This involves the economic impacts on drug development and healthcare delivery. Pharmacogenomics may impact the various dimensions of the process of drug discovery, development and marketing. These effects can result in increased or decreased development costs and drug prices with the net effect being hard to calculate without empirical research.

Category 3: Equity
This ultimately relates to the ethical decisions and consequences associated with stratified/personalized health interventions [24]. In a context of pharmacogenomically guided drug prescription and disease diagnosis, medical treatments may significantly improve for some subpopulations, but this may occur in the face of inequi-

ties because drugs may not be developed for certain groups who represent a small market for the pharmaceutical industry or a large but poor population. The existing genomics gap between developed and developing countries may potentially be exacerbated with the introduction of pharmacogenomics [29].

A point that is not discussed in-depth in the Nuffield Council on Bioethics Report is the ethics of decisions that are not made. In other words, the recognition of the notion that one is responsible not only for the decisions made (ethics of commissions) but also for decisions that are not made (ethics of omissions). When a drug is not developed for a small population that represents an inadequate financial incentive, this may remain as a silent issue as ethics of omissions tend to attract less attention.

Category 4: Control

Who should decide whether a patient should take a pharmacogenomics test? What happens if a patient does not want to undergo pharmacogenomics testing, but still wishes to access the relevant personalized medicines? Moreover, the availability of direct-to-consumer personal genomics tests call for appropriately targeted regulatory oversight. A simple extension of the regulatory risk assessment mechanisms from the genetics age (of monogenic diseases) is unlikely to be successful and in fact, might lead to underprotection of subjects and/or misdirected precaution (fig. 1). Prainsack et al. [30] aptly observed that 'a genome scan reveals information that is medical, genealogical and recreational, and those who scan and interpret the data are not distinct bodies of experts, but instead, novel configurations of geneticists, customers, ethicists, bioinformatics experts and new media executives'. There is therefore a need to rethink the outdated models of regulation in the postgenomics era [11, 30]. Increasingly blurred boundaries between experts and lay persons as well as public and private institutions in genomics research also demand a 'political science lens' to discern the motives and values that drive the direct-to-consumer availability of genomics tests in parallel to normative bioethics reasoning [11].

The above 4 categories are not mutually exclusive but overlap considerably in real-life applications of pharmacogenomics. In the recently available report on 'genomic medicine' by the Science and Technology Committee of the UK House of Lords, it is indicated that the Nuffield Council on Bioethics is due to report in 2010 on the results of its study on the ethical issues raised by new technologies that involve more personalized healthcare. While this upcoming report is not available at the time of the writing of the present book chapter, the reader is encouraged to consult with the Nuffield Council on Bioethics in 2010.

Despite the valuable points raised by the Nuffield Council on Bioethics early on pharmacogenomics, it has been suggested that it does not go far enough in acknowledging the scientific uncertainties of pharmacogenetics research or putting it into the broader context of pharmaceutical research and drug development [31]. Importantly, the Nuffield Report is based largely on anticipated future scenarios, but there is a need for further discussion of the actual contexts and the ethical and policy issues impact-

Fig. 1. Preventing misdirected precaution and designing regulatory oversight mechanisms that appropriately reflect and target the realities of postgenomics medicine. Reprinted with permission from Macmillan Publishers Ltd. [30].

ing pharmacogenomics research. One notable contentious point that is not mentioned in the Nuffield Report is the concern over DNA biobanks in the pharmaceutical industry. As noted by Corrigan [31]: 'The report fails to acknowledge the extent of the current routine practice of collecting and storing DNA samples and data during clinical drug trials. Research currently being undertaken by social scientists, such as myself, suggests that thousands of DNA samples are being collected during clinical drug trials and stored by the industry daily, and indeed most pharmaceutical company sponsored clinical drug trials now involve the collection of samples as routine.' This is consistent with the reports from the industry that DNA archives and biobanks are indeed established by the pharmaceutical companies [32]. From the broader angle of empiricism necessary in bioethics reasoning, this provides yet another example of the need to ground the bioethics debates in the practice of science and technology, so that they focus on and best reflect the actual lived experiences as discussed earlier in this chapter.

The voluntary nature of consent to participate may be compromised if pharmacogenomics testing is a condition of enrollment and when the clinical trial is the only means of having access to a particular treatment. It has been recommended that to protect the privacy of participants in research, the greatest degree of anonymity should be imposed on samples, compatible with fulfilling the objectives of the research [15]. A distinction was made between 'narrow' and 'broad' consent by the Nuffield Council Report. In the case of broad consent, it acknowledges that it may not be possible at the time of consent to specify in any detail the future use of patients' samples. The Nuffield Council Report predicts, however, that future studies will usually but not always be within the same broad areas of research as the initial project. We contend that this prediction made in 2003 is no longer tenable in the postgenomics era particularly in reference to the genome-wide association studies conducted for hypothesis generation with a systems biology orientation. There is growing interest from researchers in genomics data sharing and utilizing subjects' DNA and other biological samples for associations with a host of phenotypes that may be vastly different from the original study goals or end points that are defined a priori. These research interests and the move towards systems biology in genomics present a trade-off with individual privacy protection and protection of subjects against potential harm that may arise from data sharing or open-ended use of biological samples for exploratory research. When participants in genetic research were asked about their perspectives on DNA data sharing and the above trade-off between the scientific and clinical utility of data sharing and individual privacy protection, most participants generally wanted control over decisions about data sharing and expressed an interest in receiving information [33]. There was wide variation in their judgments about the trade-off between protecting privacy and promoting scientific and clinical utility of the data [33]. The latter study [33] underscores the need for empirical assessment of how best to define the thresholds for advancing science in the postgenomics era while ensuring protection of research subjects.

Insofar as discrimination due to disclosure and availability of genetic information is concerned, the Genetic Information Nondiscrimination Act (GINA) of 2008 extends important protection against discrimination in health insurance and employment [34]. Further fieldwork is necessary to evaluate the effectiveness of GINA in protecting individuals and populations from discrimination. Moreover, critiques note that GINA offers only a piecemeal policy solution to health inequities and risks of discrimination [35]. For example, GINA does not offer protection for other factors, such as findings on a colonoscopy that predispose individuals to illness. An insurance company can potentially discriminate against such persons in coverage and pricing because of the increased risk they represent. As we move towards predictive medicine using large-scale genomics, proteomics or other laboratory tests, it would be essential to devise policy options that broadly protect individuals who carry susceptibility markers, whether such markers are genetic or not.

A more fundamental problem emerging from large-scale genomics research and correlated genotype-phenotype datasets is whether and to what extent the anonymity, privacy or confidentiality of persons' information can be guaranteed. A technical factor that is often overlooked in previous bioethics debates is that advances in bioinformatics and genealogical analysis can re-identify individuals. A bioinformatics procedure known as k-anonymity has gained popularity to protect individuals' privacy; this was used in the Hippocratic Database Technology of the IBM [for further discussion, see 36]. It appears, however, that breaches can easily occur; this led to the development of a more robust privacy protection criterion named L-diversity [37]. Regardless of these gaps and improvements in bioinformatics technology, it is evident that absolute privacy is not fully tenable in rich genomics data sets. This neglected technical context raises the possibility that the current practices of informed consent that assure strict privacy and confidentiality may result in disingenuous consent with promises that cannot be delivered by researchers [36]. As a potential solution, an open consent procedure was proposed based on the moral principle of veracity, i.e. telling the truth. With open consent, 'volunteers consent to unrestricted redisclosure of data originating from a confidential relationship, namely their health records, and to unrestricted disclosure of information that emerges from any future research on their genotype-phenotype data set, the information content of which cannot be predicted. No promises of anonymity, privacy or confidentiality are made' [36]. Thus, a deep tension is now emerging within the field of bioethics as the scale and pace of globalized postgenomics research and medicine challenge the very notion of the 'protection paradigm' and the hitherto dominance of the autonomy principle in bioethics (and its history of development as a response to crimes against humanity or scandals over the second half of the 20th century). Whether this represents an upcoming period of identity crisis in bioethics or a reflection of growing pains as a field of inquiry lending a greater role to empiricism remains to be seen.

Race-Based Pharmacogenomics

The relationship between postgenomics medicine and the sociopolitical concepts of race and ethnicity has to be understood within the complex historical and social context. This is particularly important in psychiatry as race and ethnicity have historically played a long-standing role (and they still do) in psychopharmacology since the 1950s as a stratification axis in discerning population and individual differences in drug pharmacokinetics and pharmacodynamics. With the emergence of psychiatric genetics and pharmacogenomics, this practice, however, contrasts with the dominant view in anthropological genetics that race is a poor predictor of genotype [38]. The appropriateness of race as a variable in the present day pharmacogenomics science, and race-based therapeutics more generally, came under intensive scrutiny with the regulatory approval of BiDil for treatment of heart failure in African-Americans [39, 40]. Prior to the availability of human genomic variation data from the International HapMap Project, race has often been considered in pharmacogenomics publications as a proxy (i.e. a placeholder) for an as-yet undetermined blend of genetic, biological and environmental contributors to variability in drug treatment outcomes. Other frequently cited rationales for using race in present day practice of pharmacogenomics include the following:

- the allele frequency differences in pharmacologically significant genetic loci among human populations, e.g. the frequency of *CYP2C19* poor metabolizers in Asians (10–30%) is at least 10 times that in Caucasians (1–2%),

- the already existing use of race by physicians to improve the precision of diagnostic and therapeutic decisions, and by research sponsors, e.g. the US National Institutes of Health to ensure inclusivity in research participation,
- the unequal distribution of disease-associated alleles for certain recessive disorders such as sickle cell anemia among racially defined populations.

When race is used as a surrogate for the hereditary component of drug response, there is concern for 'conceptual drift' over it becoming a substitute for pharmacogenomics testing in prescription decisions (e.g. BiDil), while reifying the notion of race as a biological entity. Moreover, once a race-based drug is approved by regulators, there may be little incentive to carry out mechanism-oriented research by pharmaceutical manufacturers due to the risks of market shrinkage from, for example, the removal of patients who lack the genetic markers that predict a good response.

The US Office of Management and Budget Directive No. 15 (Race and Ethnic Standards for Federal Statistics and Administrative Reporting) describes the census categories used for collection of data on racially defined populations. Firmly pointing to the social construction of race, this directive emphasizes that the racial categories are not anthropologically or scientifically based [41]. This clarification made by the official US federal documents remains, however, in small print and the race-based census categories that are not scientifically based continue to be pervasively used as a proxy for genetic differences in pharmacogenomics studies. Moreover, as aptly noted by Lee [42], 'The nature of these [racial] categories is directly related to the lobbying efforts of socially identified populations rather than by scientific research that posits biologically relevant differences between individuals who subscribe to membership in any of these categories.'

Genetic admixture within and between human populations is another fundamental reason that precludes the use of race as a genetic proxy for pharmacokinetic and pharmacodynamic differences. For example, the genetic admixture in the Brazilian population firmly indicates that genetic heterogeneity in a population cannot be adequately represented by arbitrary 'race/color' categories and instead should be dealt with as a continuous variable [43]. Moreover, empirical research using tagSNPs from 3 pharmacogenetic pathways (irinotecan, 5-fluorouracil and insulin) across 270 individuals from 4 racial groups, available from the International HapMap Project, recently confirmed that race is not a major contributor to the SNP data variance. These molecular observations indicate that most genetic variation is determined by individual variation, not by racial grouping, and that race is not an adequate surrogate for individualized therapy [44]. That is, as noted by Suarez-Kurtz [43], 'each person must be treated as an individual rather than as an exemplar of a race, and that the notion of race-targeted drugs is unacceptable, especially in the case of admixed populations'.

Anthropologist Michael Montoya distinguishes the use of race as an appropriate descriptive quality to identify differences in health and health care (which addresses issues due to racism and attendant health disparities rather than race as a biological construct) from its inappropriate attributive use as a proxy for biological difference

[45]. Ellison et al. [46] suggest, however, that 'the use of these [racial] categories to promote equitable participation in biomedical research and to *describe* variation in health risk leads to the use of the same crude categories to (mis)*attribute* causality and thereby (mis)identify health care needs'.

The issue at hand is not to deny the vast differences among populations in drug efficacy and safety. Nor is the concern over race-based pharmacogenomics limited to stigma or imprecision of treatment outcomes when race is used as a predictor variable. A more fundamental problem with race concerns the impossibility of measuring or defining it as a variable, and the 'fluidity' of the concept. The following empirical research confirms and clearly exemplifies the latter point. First, running against the popular notions of race as a fixed variable, misclassifications do occur: a study of infants who died in their first year showed that 37% of infants with an entry as Native American on their birth certificates were classified as a member of another race on their death certificates [47]. Second, self-identified race and ethnicity can change over time and context. In a study conducted in the USA, one third of the study sample chose a different ethnicity when asked 2 consecutive years [48]. However, in other situations, such as individuals who face forced migration at militarized conflict zones or persecution due to human rights breaches in their native countries, people may exercise their 'exit rights' to identify with a community or ethnic group that is different than the one traditionally assigned to them by descent. Third, visual identification of race and ethnicity is notoriously inaccurate. A study that compared the ethnicity determined based on visual identification by police officers with that by analysis of the short tandem repeat loci found that those genetically assessed to be 'Middle-Eastern' were visually classified as such only about one third of the time [49]. Returning to our discussion on race-based therapeutics in the clinic, it then becomes obvious that if a patient decides to switch to a new family doctor and if each doctor visually identifies the same patient as a member of a different racial group, their 'race-based' prescriptions will also change! This can seriously jeopardize optimal therapeutics and introduce an undesirable and entirely unscientific arbitrary element in prescription decisions in the clinic. It is interesting to note that the discourse on the use of race in clinical medicine and scientific practice has thus far neglected such plasticity of human perceptions (including those of doctors) based on the visual characteristics of other human beings as well as of the variable nature of self-identified race and perceived membership in a given ethnicity. Given the rapidly escalating human migration/immigration patterns in the face of regional and international conflict zones, pandemics and globalization, it would be naive to think that individuals – as global citizens – will subscribe to a singular group identity over the course of their lifetime, nor will self-identified group memberships be perceived uniformly by others.

Debates already took place several decades ago on the (in)appropriateness of race as a stratification axis in research fields such as public health and anthropology. Due to the historically significant role given to racial differences in drug effects since the 1950s, and the present confusion within the pharmacogenomics research community on the relevance of race with the availability of more precise molecular markers of drug efficacy and safety, race-based pharmacogenomics cannot be left to random drifts in scientific practice and terminology. Lee and Mudaliar [38, 50] alerted us that 'the roughly 700 drugs in the development pipeline aimed at African-Americans signal an emerging landscape of race-based therapeutics and underline the risk of prematurely jumping from genotype to phenotype'. This calls for development of policy interventions and ethical standards both for the research sponsors and reporting of pharmacogenomics findings in postgenomics medicine. The significance of dealing with race-based pharmacogenomics practice in a timely manner becomes more obvious when considering the patients who may be already multiply marginalized and vulnerable due to mental health problems.

Personalized Medicine beyond Pharmacogenomics: Approaching Wave of 'Proteomics Diagnostics' in Psychiatry and Potential for Disease Mongering

Until the 18th century, individuals had to feel and demonstrate florid (fully developed) physical findings to be considered as 'sick'. With the introduction of personalized medicine, the concept of what is considered as being sick is changing rapidly: there is a growing emphasis on prediction of future disease liabilities and individual health risks. One can now be considered as sick or a 'future patient' based on biomarker tests and measurements, in the absence of physical symptoms. A traditional example is the plasma cholesterol measurements. Personal genomics, too, is influencing how common complex disease risks are being framed particularly with the introduction of direct-to-consumer whole-genome testing. Personalized medicine is thus blurring the boundaries of treatable diseases, and causing a 'temporal shift' in the folklore of modern medicine and therapeutics – from present to the future. Expanding the limits of what is treatable would make sense in a context of preventive medicine and early clinical intervention, especially when a given disease is at its prodromal and preclinical stage. However, this also requires that diagnostic tests for personalized risk assessment are available with acceptable analytical validity and clinical utility.

The term 'disease mongering' refers to artificially expanding or inappropriately inflating the limits of what is treatable, e.g. as a result of commercial motivations to increase the sales of health products such as pharmaceuticals. Concerns around disease mongering have been expressed in psychiatry [51]. Conceivably, the above shifts in diagnostic medicine including direct-to-consumer personal genomic tests may have profound effects in the case of clinical psychiatry where persons' subjective experiences and symptoms may be confounded or shaped by diagnostic tests that lack an appropriate evidentiary base.

Proteomics is a next-generation high-throughput 'omics technology' that aims to characterize functional variability in cell function in health and disease [for a review, see 52]. Proteomics has a notable difference in contrast to genotype-based diagnostics. Proteomics brings about a more dynamic form of ongoing diagnostic testing within the same individual, to obtain a longitudinal 'repeated measures' functional risk signature – as opposed to between-patient cross-sectional static point estimates of risks estimated by genotype-based tests. This anticipated further shift in conceptualization of 'health risks' brought about by proteomics and advances in postgenomics science and technology may have substantial impacts on psychiatric disease nosology, boundaries between normal and deviation from normalcy, and what may be considered as a treatable illness or symptom. With proteomics diagnostics, and pharmacoproteomics more specifically (i.e. study of human proteome variation in relation to drug efficacy and safety) [52], disease mongering will need further reflection in clinical psychiatry.

Concluding Remarks

Postgenomics medicine is characterized by unprecedented opportunities to strengthen the healthcare systems and services as well as formidable challenges, for example, an exponential increase in the pace and scale of biomedical research, proliferation of interest groups with diverse interests that are often competing and conflicting, direct-to-consumer personal genome testing, genome-wide association studies that demand recruitment of large cohorts of subjects from diverse populations, a globalized science facing local and regional nuances in scientific and bioethics practice, to name a few. Past lessons from the genetically modified organisms and other innovations in the 20th century tell us that biotechnologies have both intended and unintended effects on science, medicine and society. If these effects are not carefully tended and monitored, they can significantly impede the sustainable and equitable development and uptake of innovations. With the approaching wave of pharmacoproteomics, systems biology and other more 'dynamic' measures of cell function, complexities and opportunities in postgenomics medicine will continue to grow. Hence, within the limits of a concise book chapter, we aimed to highlight why we need to critically examine our long held views and practices not only in pharmacogenomics but also in bioethics. Regulatory oversight mechanisms and bioethics frameworks from the genetics age (of monogenic diseases) are not well equipped to address the complexities and nuances of postgenomics medicine. As noted in the second section of this chapter, there is a need for empirical grounding of bioethics reasoning. Traditionally, such calls have been made to integrate social sciences, such as sociology and anthropology, to provide an empirical context for traditional philosophical approaches to bioethics. However, the present chapter proposes that while genomics scientists have long been the 'subjects' of bioethics analyses in the past, they can actively take part in the process of bioethics research by identifying the fault lines in scientific practice that have bioethics significance; thus, offering a hitherto neglected empirical context for both philosophers and social scientists engaged in bioethics.

Still, such calls for transdisciplinary engagement are easier said than done. The Norwegian-American sociologist and economist Thorstein Veblen referred to 'trained incapacity' [53] where one's own disciplinary expertise results in blind spots to perceive the strengths of other disciplines or 'ways of knowing', a human condition that is applicable to all of us. Yet neglecting the need to focus bioethics analyses and reasoning within scientific, technical, social, political and other contexts strips the practice of bioethics from the lived realities of science and medicine, nor does it leave room for effective policy interventions.

On his trial for heresy for encouraging his students to challenge the accepted beliefs of the time, Socrates said, 'the unexamined life is not worth living'. This notion applies to the modern day practice of individual disciplines and professional expertise as well. At this critical juncture when pharmacogenomics and related biotechnology innovations are beginning to diffuse into the 'postgenomics clinic' in psychiatry, it is time to reexamine the old and new social, ethical and policy issues [54–56], but with due attention to the decision-making processes and methodologies used to arrive at conclusions in both pharmacogenomics and bioethics [57]. This evidence-based approach is ultimately crucial to design effective and appropriately targeted real-life bottom-up (instead of top-down) policy interventions when the bioethics theory and the practice of science and medicine are not aligned.

Acknowledgments

The views expressed in this chapter are the personal ideas and interpretations of the author, and do not represent the opinions of the affiliated institutions. Supported by an operating ethics catalyst grant (R0016158) from the Canadian Institutes of Health Research, and a career investigator salary for science and society research in personalized medicine from the Fonds de la recherche en santé du Québec to V.O.

References

1 Bertilsson L, Palmér L: Indole-3-acetic acid in human cerebrospinal fluid: identification and quantification by mass fragmentography. Science 1972; 177:74–76.
2 Eichelbaum M, Bertilsson L: Determination of carbamazepine and its epoxide metabolite in plasma by high-speed liquid chromatography. J Chromatogr 1975;103:135–140.
3 Gram LF, Christiansen J: First-pass metabolism of imipramine in man. Clin Pharmacol Ther 1975;17: 555–563.
4 De Leon J: Pharmacogenomics: the promise of personalized medicine for CNS disorders. Neuropsychopharmacology 2009;34:159–172.
5 Human Genome Project: Ethical, Legal and Social Issues. www.ornl.gov/sci/techresources/Human_Genome/elsi/elsi.shtml (accessed November 9, 2009).
6 Collins FS, Green ED, Guttmacher AE, Guyer MS; US National Human Genome Research Institute: A vision for the future of genomics research. Nature 2003;422:835–847.
7 ELSA, European Commission: Ethical, Legal and Social Aspects of the Life Science and the Technologies. http://cordis.europa.eu/elsa-fp4/home.html (accessed November 9, 2009).
8 Hedgecoe AM: Critical bioethics: beyond the social science critique of applied ethics. Bioethics 2004; 18:120–143.

9 Illinois White Paper: Improving the System for Protecting Human Subjects: Counteracting IRB 'Mission Creep'. Center for Advanced Study, University of Illinois. 2005.
10 Guston DH: Innovation policy: not just a jumbo shrimp. Nature 2008;454:940–941.
11 Ozdemir V, Suarez-Kurtz G, Stenne R, Somogyi A, Someya T, Kayaalp SO, Kolker E: Risk assessment and communication tools for genotype associations with multifactorial phenotypes: the concept of 'edge effect' and cultivating an ethical bridge between omics innovations and society. OMICS 2009; 13:43–62.
12 Ozdemir V: What to do when the risk environment is rapidly shifting and heterogeneous? Anticipatory governance and real time assessment of social risks in multiply marginalized populations can prevent ethical inflation or underestimation of risks. Am J Bioethics 2009;9:65–68.
13 Egalité N, Ozdemir V, Godard B: Pharmacogenomics research involving racial classification: qualitative research findings on researchers' views, perceptions and attitudes towards socioethical responsibilities. Pharmacogenomics 2007;8:1115–1126.
14 World Health Organization: Genomics and World Health. Geneva, WHO, 2002. www.who.int/rpc/genomics_report.pdf (accessed November 9, 2009).
15 Pharmacogenetics: Ethical Issues. London, Nuffield Council on Bioethics, 2003.
16 Advisory Committee on Genetics, Health, and Society: Realizing the Potential of Pharmacogenomics: Opportunities and Challenges; A Report of the Secretary's Advisory Committee on Genetics, Health, and Society. Health and Human Services, 2008, p 24. http://oba.od.nih.gov/oba/SACGHS/reports/SACGHS_PGx_report.pdf.
17 House of Lords Science and Technology Committee: 2nd Report of Session 2008–09. Genomic Medicine Volume I. London, Authority of the House of Lords, 2009.
18 Nuremberg Code: Trials of war criminals before the Nuremberg Military Tribunals under Control Council Law. Washington, US Government Printing Office, 1949, pp 181–182.
19 Belmont Report: Ethical Principles and Guidelines for the Protection of Human Subjects of Research (DHEW Publication No. (OS) 78–0012). National Commission for the Protection of Human Subjects in Biomedical and Behavioral Research. 1978.
20 Beauchamp TL, Childress JF: Principles of Biomedical Ethics, ed 6. Oxford, Oxford University Press, 2008.
21 Fox RC: More than bioethics. Hastings Cent Rep 1996;26:5–7.
22 Fox RC, Swazey JP: Examining American bioethics: its problems and prospects. Camb Q Health Ethics 2005;14:361–373.
23 Guessous I, Gwinn M, Yu W, Yeh J, Clyne M, Khoury MJ: Trends in pharmacogenomic epidemiology: 2001–2007. Public Health Genomics 2009; 12:142–148.
24 Ozdemir V, Someya T: A transdisciplinary forum for study of individual and population variability in response to health interventions and personalized medicine (editorial). Curr Pharmacogenomics Person Med 2009;7:146–148.
25 Corrigan O: Empty ethics: the problem with informed consent. Sociol Health Illn 2003;25:768–792.
26 Burton B: Proposed genetic database on Tongans opposed. BMJ 2002;324:443.
27 Hopkins MM, Ibarreta D, Gaisser S, Enzing CM, Ryan J, Martin PA, Lewis G, Detmar S, van den Akker-van Marle ME, Hedgecoe AM, Nightingale P, Dreiling M, Hartig KJ, Vullings W, Forde T: Putting pharmacogenetics into practice. Nat Biotechnol 2006;24:403–410.
28 UK Pharmacogenetics Study Group: Policy Issues in Pharmacogenetics: A Policy Briefing from the UK Pharmacogenetics Study Group. 2006. www.york.ac.uk/res/pgx/publications/PGxpolicyissues2006.pdf (accessed November 9, 2009).
29 Pang T: Pharmacogenomics and personalized medicine for the developing world – too soon or just-in-time? A personal view from the World Health Organization. Curr Pharmacogenomics Person Med 2009;7:149–157.
30 Prainsack B, Reardon J, Hindmarsh R, Gottweis H, Naue U, Lunshof JE: Personal genomes: misdirected precaution. Nature 2008;456:34–35.
31 Corrigan OP: Pharmacogenetics, ethical issues: review of the Nuffield Council on Bioethics Report. J Med Ethics 2005;31:144–148.
32 Williams JA, Andersson T, Andersson TB, Blanchard R, Behm MO, Cohen N, Edeki T, Franc M, Hillgren KM, Johnson KJ, Katz DA, Milton MN, Murray BP, Polli JW, Ricci D, Shipley LA, Vangala S, Wrighton SA: PhRMA white paper on ADME pharmacogenomics. J Clin Pharmacol 2008;48: 849–889.
33 McGuire AL, Hamilton JA, Lunstroth R, McCullough LB, Goldman A: DNA data sharing: research participants' perspectives. Genet Med 2008; 10:46–53.
34 Genetic Information Nondiscrimination Act of 2008, Pub. L. 110–233, 122 Stat. 881, 2008.
35 Korobkin R, Rajkumar R: The Genetic Information Nondiscrimination Act – a half-step toward risk sharing. N Engl J Med 2008;359:335–337.
36 Lunshof JE: From genetic privacy to open consent. Person Med 2006;3:187–194.

37 Machanavajjhala A, Kifer D, Gehrke J, Venkitasubramaniam M: *L*-diversity: privacy beyond *k*-anonymity. ACM Trans Knowl Discov 2007;1:article 3.
38 Lee SS, Mudaliar A: Medicine. Racing forward: the Genomics and Personalized Medicine Act. Science 2009;323:342.
39 Bloche MG: Race-based therapeutics. N Engl J Med 2004;351:2035–2037.
40 Kahn J: Exploiting race in drug development: BiDil's interim model of pharmacogenomics. Soc Stud Sci 2008;38:737–758.
41 US Office of Management and Budget: Standards for Maintaining, Collecting, and Presenting Federal Data on Race and Ethnicity. 1997. www.whitehouse.gov/omb/inforeg/r&e_app-a-update.pdf (accessed November 9, 2009).
42 Lee SS: The ethical implications of stratifying by race in pharmacogenomics. Clin Pharmacol Ther 2007;81:122–125.
43 Suarez-Kurtz G: Pharmacogenomics and the genetic diversity of the Brazilian population. Cad Saude Publica 2009;25:1650–1651.
44 Yen-Revollo JL, Auman JT, McLeod HL: Race does not explain genetic heterogeneity in pharmacogenomic pathways. Pharmacogenomics 2008;9:1639–1645.
45 Montoya MJ: Bioethnic conscription: genes, race, and mexicana/o ethnicity in diabetes research. Cult Anthropol 2007;22:94–128.
46 Ellison GTH, Smart A, Tutton R, Outram SM, Ashcroft R, Martin P: Racial categories in medicine: a failure of evidence-based practice? PLoS Med 2007;4:1434–1436.
47 Hahn R, Mulinare J, Teutsch S: Inconsistencies in coding race and ethnicity between birth and death in US infants. JAMA 1992;267:259–263.
48 Leech K: Runnymede Research Report. London, Runnymede Trust, 1989. A Question in Dispute: The Debate About an 'Ethnic' Question in the Census.
49 Lowe A, Urquhart A, Foreman L, Evett I: Inferring ethnic origin by means of an STR profile. Forensic Sci Int 2001;119:17–22.
50 Medicines in Development for Major Diseases Affecting African Americans. Washington, Pharmaceutical Research and Manufacturers Association of America, 2007.
51 Moncrieff J, Thomas P: The pharmaceutical industry and disease mongering: psychiatry should not accept so much commercial sponsorship. BMJ 2002;325:216.
52 Paik YK, Graham JE, Jones M, McDonald F, Holmes C, Ozdemir V: Interview with Dr. Young-Ki Paik, President of the Human Proteome Organization (HUPO): pharmacoproteomics and the approaching wave of 'proteomics diagnostics'. Curr Pharmacogenomics Person Med 2009;7:243–248.
53 Veblen T: The Instinct of Workmanship and the State of the Industrial Arts. New York, Macmillian, 1914.
54 Ozdemir V, Husereau D, Hyland S, Samper S, Salleh MZ: Personalized medicine beyond genomics: new technologies, global health diplomacy and anticipatory governance. Curr Pharmacogenomics Person Med 2009;7:225–230.
55 Knoppers BM: Genomics and policy-making: from static models to complex systems? Hum Genet 2009;125:375–379.
56 Kato K, Kano K, Shirai T: Science communication: significance for genome-based personalized medicine – a view from the Asia-Pacific. Curr Pharmacogenomics Person Med 2010;8:92–96.
57 Ozdemir V, Knoppers BM: One size does not fit all: towards 'upstream ethics'? Am J Bioethics 2010; 10:42–44.

Vural Ozdemir, MD, PhD, DABCP
Department of Social and Preventive Medicine, School of Public Health, University of Montreal
6128, Succ. Centre-Ville
Montreal, QC H3C 3J7 (Canada)
Tel. +1 514 343 6111, Fax +1 514 398 8954, E-Mail vural.ozdemir@umontreal.ca

Schwab M, Kaschka WP, Spina E (eds): Pharmacogenomics in Psychiatry.
Adv Biol Psychiatry. Basel, Karger, 2010, vol 25, pp 30–45

Pharmacogenomics and Personality: Role of CYP2D6 and Implications for Psychopathology

Eva M. Peñas-LLedó[a] · Pedro Dorado[a, b] · Adrián LLerena[a, b] ·

[a]CICAB Clinical Research Centre, Extremadura University Hospital, and [b]Department of Medical and Surgical Therapeutics, Faculty of Medicine, University of Extremadura, Badajoz, Spain

Abstract

CYP2D6 polymorphism is related to absent (poor metabolizers, PMs) and normal/decreased (extensive metabolizers, including ultrarapid metabolizers) metabolism of clinically important drugs. Moreover CYP2D6 is also involved in the metabolism of endogenous substrates in the brain which appear highly important for psychological functioning that may be related to normal and abnormal behavioral tendencies or psychopathology. Although there are contradictory results, a lot of support has been given to the role of CYP2D6 activity on behavioral and mental functions. PMs have been consistently associated with a personality profile characterized by impulsive- and anxiety-related features. Despite these psychological features often being linked to psychopathology, there also appears to be evidence suggesting that PMs could present some protective features against the development of mental disorders. In keeping with this notion, PMs were found to perform better in a sustained attention cognitive task, which is impaired in different syndromes characterized by high vulnerability to stress and impulsivity such as schizophrenia or attention-deficit hyperactivity disorder, etc. In conclusion, there are data supporting the view that CYP2D6 may influence not only variability of drug response but also psychological functioning, and hence may be related to vulnerability to psychopathology.

Cytochrome P450 CYP2D6 is one of the most important enzymes involved in the metabolism of a large number of drugs used worldwide, although its role in endogenous metabolism has not been fully clarified [1, 2]. Multiple allelic variants of the *CYP2D6* gene have been identified which are associated with absent or increased enzyme activity in individuals who are termed, respectively, poor metabolizers (PMs) and extensive metabolizers (EMs), including ultrarapid metabolizers (UMs) [3]. The present review is an update of research on the impact of CYP2D6 polymorphism on 'personality traits'. We shall first review evidence about the location of CYP2D6 in the brain, as well as its potential impact on the endogenous metabolism of neurotransmitters and hormones. Then we shall discuss the relationship of CYP2D6 with 'personality' and psychopathology.

CYP2D6 in the Brain

In general, investigations of the localization, expression and function of CYP2D have mostly focused on the rodent brain. Six isoforms of the CYP2D subfamily have been identified in rats, namely CYP2D1 through CYP2D5 and CYP2D18 [4]. Among them, CYP2D4 and CYP2D1 (the gene corresponding to human CYP2D6) are thought to be fairly abundant in the CNS [5–7]. It has been demonstrated that CYP2D1 mRNA is widely distributed and constitutively expressed in neuronal and glial cells in diverse brain areas such as the cerebral cortex, hippocampus, dentate gyrus, caudate, putamen, hypothalamus and thalamus [8]. Conversely, a more recent study failed to detect CYP2D1 in the rat brain using immunohistochemical techniques [9].

Knowledge of CYP2D6 in the human brain is still limited. An investigation showed metabolism of dextromethorphan to dextrorphan in human brain microsomes [10]. CYPD6 protein and messenger RNA expression appear to be specific to neurons in the human cerebral cortex, hippocampus and cerebellum [11]. CYP2D6 was first identified in brain tissues by immunoblotting [12]. Additional evidence was provided by McFadyen et al. [13] showing CYP2D6 expression in midbrain. In particular, Gilham et al. [14] revealed a more precise localization of CYP2D6 within this brain area by using in situ hybridization, and showed that pigmented neurons corresponded to the substantia nigra, which plays an important role in movement and in reward seeking and learning.

In another study, Siegle et al. [15] analyzed the regional and cellular expression of CYP2D6 transcripts and proteins in postmortem brain tissues of 3 individuals. A combination of in situ hybridization coupled with immunohistochemistry on adjacent sections allowed simultaneous detection of CYP2D6 mRNA and protein. However, discrepancies existed in the results indicating that the mRNA was more widely distributed in the brain areas analyzed compared to the protein. Neuronal cells, as well as glial cells, showed labeling for mRNA in such brain regions as the substantia nigra, caudate nucleus, putamen, globus pallidus (nuclei of the basal ganglia rich in dopaminergic neurons), hippocampus, hypothalamus, thalamus, cerebellum and neocortex. In contrast, CYP2D6 protein was primarily localized in large principal neurons, such as pyramidal cells of the hippocampus and cortex, and Purkinje cells of the cerebellum. In glial cells, CYP2D6 protein was absent.

A different study examined the regional and cellular expression of CYP2D6 mRNA and protein by RT-PCR, Southern blotting, slot blotting, immunoblotting and immunocytochemistry in humans. A significant correlation was found between mean mRNA and CYP2D6 protein levels across 13 brain regions. In hippocampus, this was localized in CA1, CA2 and CA3 pyramidal cells, and dentate gyrus granular neurons. In cerebellum, it was localized in Purkinje cells and their dendrites [11]. The above studies provide evidence of CYP2D6 expression in certain regions of the CNS and may indicate a role of CYP2D6 in specific psychological functions.

CYP2D6 and Endobiotic Metabolism

The CYP enzymes, beyond their contribution to detoxifying foreign chemicals such as xenobiotics and drugs, are involved in the biotransformation of endobiotics. They are also involved in the metabolism of endogenous substances, which may influence both normal and altered physiological processes. In our earlier study on the resulting association between CYP2D6 and some personality traits (see below), we suggested a role of CYP2D6 in the metabolism of endogenous psychoactive substrates or products, hypothesizing its participation in the metabolism of a neuroactive amine [16].

Different studies have been performed to elucidate the role of CYP2D6 in endogenous metabolism. In vitro studies have found that CYP2D6 is involved in the biotransformation of tyramine to dopamine (DA) [6, 17]. CYP2D6 has been also found to be involved in the O-demethylation of the β-carbolines harmaline and harmine [18] and the regeneration of serotonin (5-HT) from 5-methoxytryptamine [18–20]. These studies suggest the potential influence of CYP2D6 polymorphism in the balanced functioning of the DA and 5-HT endogenous systems. In keeping with this, Kirchheiner et al. [21] tested whether PMs presented lower 5-HT concentrations in platelets than UMs and EMs at 2 different moments, and found positive results. Also, Ozdemir et al. [22] presented evidence for an explanatory model on the potential relationships between CYP2D6 polymorphism and 5-HT and DA activity, where PMs would present lower 5-HT levels but higher DA tone in the pituitary [23]. Interestingly, alterations in 5-HT receptors or transporters in the brain causing a hyposerotonergic tone have been shown in humans to contribute to deficits in affect regulation that can predispose to anxiety or stress-related disorders [24] or to different aspects of impulsivity [25]. Therefore, the implication of CYP2D6 in personality could be mediated by the influence of this enzyme activity in the serotonergic tone.

CYP2D6 is also implicated in the metabolism of other endogenous compounds such as anandamide, the endogenous ligand to the cannabinoid receptor CB1 [26]. Other studies suggest the possibility that CYP2D6 may be involved in the regulation of endogenous neuroactive steroids such as progesterone and its derivatives in brain tissues [27, 28], which may also play a role in behavioral processes and psychological traits.

CYP2D6 and Psychological Functioning: Relevance of Personality Traits

CYP2D6 and Personality: Earlier Studies and Hypotheses

The first hypothesis postulating that CYP2D6 could have an endogenous neuroactive substrate or product such as a biogenic neurotransmitter amine was put forward in 1993 by Llerena et al. [16]. After a series of studies, they presented the first evidence for the potential impact of CYP2D6 on personality on the basis of preliminary observations about the idiosyncratic personality profile of PMs by Bertilsson et al. [29]. This was the first study using a psychological measure to establish differences between PMs and EMs. In a population of healthy Swedish individuals who were also studied phenotypically, PMs were reported as presenting lower psychasthenia than EMs [29]. Contemporaneously, Llerena et al. [30] found a greater frequency of CYP2D6 PMs among unrelated volunteers involved in a clinical trial. Later, Llerena with Bertilsson, Shalling (who develop the Karolinska Scales of Personality, KSP, used to evaluate personality) at the Karolinska Institute in Sweden, and other collaborators analyzed whether these differences were replicated in a large and independent population of unrelated healthy volunteers from the University of Extremadura in Spain [16]. To this end, they compared previously phenotyped PMs with different groups of EMs classified according to their CYP2D6 debrisoquine hydroxylation activity on the same psychological measure (KSP) which, by then, had already been translated into Spanish [31, 32]. Again, PMs were shown to present differences with respect to all groups of EMs. PMs in this population presented higher levels of psychic anxiety and lower levels of socialization than all EM groups.

A second hypothesis based on the role of CYP2D6 in endogenous 5-HT metabolism [19] postulated the existence of a hyposerotonergic/hyperdopaminergic tone in PMs [22, 23, 33]. This proposal may well help to clarify the interpretation of findings about the CYP2D6-personality relationship. A balance characterized by low levels of 5-HT and as a consequence by high levels of DA has been related to a cluster of behavioral traits, including impulsivity and anxiety, that have been found in CYP2D6 PMs [16, 34, 35]. Nevertheless, the relationship between CYP2D6 and personality in different populations and cultures has yielded results associated either with low or high levels of anxiety- or with impulsivity-related traits. This suggests that other important factors may be influencing the expression/direction of personality traits and therefore may bias this relationship. Among such factors the most important one appears to be illness. In keeping with this, we have separated the studies according to whether they were carried out on healthy or clinical samples to better discuss their results in the light of current hypotheses.

The original studies on the role of CYP2D6 in metabolizing endogenous substrates and the potential effects on behavior motivated a line of research that continues to the present. However, the apparent conflictive nature of some of these findings and the new ones has, on occasions, led more to confusion rather than to understanding. We will review most of the studies designed to explore the relationship between CYP2D6 and personality with the aim of clarifying the existing data.

CYP2D6 and Personality Studies in Healthy Volunteers

As mentioned above, several studies have compared CYP2D6 PMs versus EMs in healthy volunteer populations. Three of those studies used the same phenotyping method as determined by debrisoquine hydroxylation capacity, and the same personality measure (KSP) in Sweden, Spain, and Cuba [16, 29, 34, 35]. This relationship has also been analyzed in healthy volunteers from Germany and Japan. However, those studies were characterized by using a different methodological approach for CYP2D6 activity and personality [36–38].

In the first study in Sweden [29], PM participants presented lower 'psychasthenia', a trait that is associated with lower fatigability, greater motivation and greater emotion communication. Its measure is clearly linked to approach-focused behaviors instead of to approach-avoidance ones, which is why some authors have related it to 'novelty seeking' or impulsivity [22, 35]. However, in the other 2 studies that followed the same methodological approach, the results differed. In Spain [16], PMs versus another 3 groups of EMs (classified according to their CYP2D6 activity) were found to score higher in 'psychic anxiety' and impulsivity-related traits such as 'low socialization'. In the third study in Cuba [34], designed to replicate the Spanish one, the same results of 'psychic anxiety' and 'low socialization' were found among PMs. The fact that these last 2 studies found the same pattern of responses appeared to point to a biological phenomenon: that CYP2D6 variability relates to behavior regardless of the environment. They were carried out in 2 different countries, with a time lag between them of more than a decade, at different latitudes, and with different economic and social development. However, there were other factors that were clearly shared by the 2 populations: a Latin background, language, cultural and historical factors, education, age, etc. Also, the 2 studies used the same design and methodological approach. On the contrary, the Swedish population appeared more heterogeneous and with a greater mean age. This divergence in characteristics with regard to the sample composition and the differences in cultural and other aspects may have made participants interpret the items to complete in the self-report questionnaire differently in Cuba and Spain versus Sweden.

The relationship between CYP2D6 and personality in healthy volunteers has also been analyzed in Germany and Japan. These studies were characterized by using a different methodological approach. Firstly, they did not analyze the phenotype by administering a probe drug such as debrisoquine or dextromethorphan. Instead, they estimated the phenotype by genotyping. Secondly, they used different personality measures, which makes comparisons across studies even more difficult. Thirdly, in Germany, PM versus EM individuals were recruited in the community by newspaper advertisement [36]. In that study, differences were only found between PM and EM women. PMs were characterized by greater 'consciousness' in the NEO Personality Inventory (NEO-PI) [39], which measures responsibility, orderliness and an aim for achievement through perseverance. This trait has been consistently associated with hard-working, perfectionist and reliable individuals, who show good performance across all job categories in several meta-analyses combining all kinds of studies using the NEO-PI in different continents [40]. Indeed, the perfectionist and hard-working characteristics measured by the 'consciousness' scale of the NEO-PI are personality features linked to the functional or efficient effects of optimum anxiety levels, whereas the pursuit of unrealistic standards would be linked to a lack of consciousness and to affective and anxiety disorders. Therefore, all the above results may be interpreted under the common label of functional psychic anxiety or perfectionism that may allow PMs to anticipate all potential outcomes in order to prepare themselves with the result of increased performance and efficiency.

In Japan, 2 other studies on healthy volunteers [37, 38] also determined only the *CYP2D6* genotype, and personality was measured with the Temperament and Character Inventory (TCI). Neither study found any association between CYP2D6 variability and personality. The main problem with these studies was the very low number of PMs together with the very little CYP2D6 variability. Indeed, in the first study by Suzuki et al. [37], there were no PMs given that they only analyzed differences between those individuals homozygous or heterozygous for the *CYP2D6*10* variant (with decreased but not null activity) and those with functional or wild-type alleles. Therefore, no differences could be said to be found between EMs (**1/*1*) and individuals with **10/*10* and **1/*10* genotypes. However, those with reduced CYP2D6 activity appeared to present a tendency to higher impulsivity (novelty seeking).

Similarly, in the second study [38], PMs were represented by just 1 subject with 2 null alleles (**5/*5*). This PM was compared with the remaining 341 participants who were separated into 3 groups of: 'intermediate' metabolizers with either 2 reduced activity alleles or 1 reduced activity allele and 1 non-activity allele (e.g. **4/*10, *5/*10, *10/*10*), EMs with 1 or more wild-type alleles (e.g. **1/*1, *1/*2, *1/*5, *1/*10*), and UMs considered to be those with more than 2 copies of wild-type (e.g. **1/*1XN, *1/*2XN*). Despite there only being one PM (n = 1), this individual presented a higher score on impulsivity (novelty seeking) than the mean, minimum, and maximum scores found within EMs, intermediate metabolizers, and UMs, with the UMs being the group (n = 4) that scored lower than the rest in this dimension. This PM also scored lower on harm avoidance (a dimension considered a marker of depression and other affective disorders) than the other groups' scores (minimum and maximum) in this dimension.

Controversy in Personality Studies in Healthy Volunteers

As noted above, personality trait studies in CYP2D6 PM versus EM healthy volunteers have yielded mixed results [16, 29, 34]. While in the first study in Sweden [29], PM participants presented lower 'psychasthenia' (associated with lower fatigability and greater motivation and emotion communication), PMs from Spain [16] and Cuba [34] scored higher in 'psychic anxiety' and impulsivity-related traits such as 'low socialization'. The sources of variability in the CYP2D6 activity-personality relationship in these healthy volunteer studies might be due to recruitment procedures (as discussed above) and the participants' differences in education or other environmental factors including culture. The studies in Spain and Cuba used the same methodological approach, and identical recruitment procedure and participant type (university students and staff in hospitals and medical schools, of about the same age). However, the recruitment in Sweden was different, and the population was more heterogeneous and older. Thus, the characteristics of healthy volunteers related to cultural, educational and recruitment factors, among others, might be linked to a population selection bias, which may be the cause of differing results across studies in the relationship between personality and CYP2D6.

Differences found in the other studies analyzing the relationship between CYP2D6 genetic polymorphism and personality in Germany [36] and Japan [37, 38] may additionally be related to different sources of variability related to the study's methodological approach. Firstly, the evaluation of CYP2D6 hydroxylation capacity was extrapolated from the genotype, since the actual enzyme hydroxylation capacity (phenotype) was not studied. Therefore, differences in the relationship between CYP2D6 pheno- and *CYP2D6* genotype may affect the results. Secondly, differences in personality evaluation may occur since NEO and TCI were used in Germany and Japan, respectively, whereas KSP was used in Sweden, Spain, and Cuba. Thirdly, there were the differences in the kinds of participants who were recruited.

In conclusion the sources of variability in the CYP2D6 activity-personality relationship in healthy volunteer studies might be due to participant differences in factors such as age, sex, ethnic background, culture, and education due to recruitment procedures and population characteristics. Differences may also be due to CYP2D6 and personality evaluation procedures. Furthermore, there might also be a personality-linked population selection bias concerning all healthy volunteers participating in biomedical studies that involve an invasive procedure as well as responding to a psychological measure protocol. This self-selection bias aspect will be analyzed in more detail below.

CYP2D6 and Personality Studies in Clinical Settings

There are 2 studies that have evaluated personality differences between PMs and EMs detected in clinical populations. The first of these was carried out on 'healthy' Malaysian individuals who were in hospital to undergo orthopedic surgery [41]. These Asian participants were compared regarding personality on the basis of their genotypes. Again, no PMs were detected, but the researchers clustered all those with null or reduced activity alleles under the label of 'slow' (**4/*10; *5/*10; *10/*10; *10/*17*), and those with at least 1 null or reduced activity allele under the label of 'intermediate' (**1/*4, *1/*5; *1/*9; *1/*10*), and compared them with those that they considered normal (**1/*1*). Participants were evaluated on a measure of 'type A personality', a behavioral pattern characterized by tenseness, impatience, urgency and aggressiveness, which often results in stress-related disorders. Both groups including participants carrying alleles with reduced or null activity (the 'slow' and 'intermediate' groups) scored lower on 'type A personality' or on the vulnerability to develop a stress-induced disorder [41].

The other study included patients suffering from major depressive disorder in New Zealand [42]. PMs in this group (n = 8) versus EMs (n = 113) were found to score significantly lower on anxiety-related traits. These genetically determined *CYP2D6* PMs versus EMs scored lower on the TCI-R 'harm avoidance' and all its subscales, 'fear of uncertainty', 'fatigability' and 'shyness', with the exception of 'worry/pessimism', a trait closely related to psychic anxiety and perfectionism [42]. Curiously, the overall score on these 'harm avoidance' scales, in which PMs scored lower, is proposed as an endophenotype or marker for the development of depression. Furthermore, these PMs scored higher on impulsivity related traits (novelty seeking score), although this association did not reach significance. These data seemed to suggest that PMs recruited in clinical samples or settings, who are exposed to more stressful conditions, report personality features that are often negatively associated with severity and duration of psychopathology [43]. Since these studies did not report on the history of psychopathology or clinical evolution, it is difficult to draw conclusions about the potential effects for the development of pathology in PMs.

CYP2D6 and Personality Studies: Population Selection and Evaluation Bias

The aforementioned differences in the relationship between CYP2D6 and personality may be due to a population selection bias, where self-proclaimed healthy volunteers may include not only healthy but also 'clinical', 'subclinical' or individuals vulnerable to psychopathology, or experiencing episodic distress. It is known that a percentage of subjects participating in studies including personality measures present disordered personality features related to mental disorders [44–50]. This bias may be reduced by including a psychopathological screening of the participating healthy volunteers.

Therefore, in order to evaluate this potential bias, we have analyzed the relationship between *CYP2D6* and personality in healthy volunteers, controlling for their history of psychopathology and psychological status as explained in the following [35]. Firstly, participants underwent a general psychiatric interview in order to detect history of any psychiatric problem or psychotropic treatment (psychopathology). Secondly, once we selected those free of psychiatric disorders, general episodic distress was evaluated with the Symptom Checklist-90-Revised (SCL-90R) [51] at the moment of the evaluation. Personality was evaluated with 2 of the most frequently used measures, the KSP and TCI [35].

Consistent with the aforementioned studies, a group of participants who considered themselves healthy volunteers were found to present a history of psychiatric disorders or psychopharmacological treatment. Thus, it appears that if healthy volunteer studies do not include a psychiatric interview, disordered people may be considered to be healthy, thus biasing the results.

Furthermore, when psychological status was evaluated in those without psychiatric problems, interestingly all PMs scored below the cutoff score suggestive of distress ('positive affect'). However, EMs were found both below and above the cutoff point ('positive' and 'negative' affect, respectively). This is important because experiencing positive or negative mood at the moment of evaluation may also influence the evaluation of personality. In order to evaluate the potential relevance of psychological distress for assessing personality, KSP and TCI were compared between subjects with 'positive' versus 'negative' affect within the EM group. Most KSP subscales (with the exception of monotony avoidance and impulsivity) and TCI scales (with the exception of novelty seeking and reward dependence) were different between those groups [35]. These results clearly indicate that general distress at the moment of evaluation may bias the resulting relationship between CYP2D6 and personality.

These findings appear also to be pertinent for another potential bias about an over-representation of certain personality features among participants in biomedical research. It has been reported that they usually present higher than normative scores on impulsivity and related features (novelty or sensation seeking, monotony avoidance, etc.) [52, 53]. In keeping with that, we found an over-representation of CYP2D6 among healthy volunteers participating in a clinical trial [30].

To further explore how distress at the moment of personality evaluation could bias personality results, we decided to compare the mean KSP scores obtained for the whole population in our 2 previous studies in Spain [16] and Cuba [34] with mean KSP personality scores obtained in the groups with distress and without distress from this population [35]. We found that there were no differences in any subscale between mean KSP scores from the study in Spain [16], Cuba [34] and the group with distress of the last Spanish study [35]. Again, consistently with the 'impulsivity' bias found in biomedical research participants, the above 3 groups presented differences in all KSP personality scales from the positive affect group in the last Spanish study [35], with the exception of the scales of 'impulsivity' and 'monotony avoidance'.

Personality Characteristics of PMs: Current Knowledge and Scientific Bases

PM Personality in Healthy Volunteers

In our latest study [35], we found that genotypically inferred PMs scored higher than EMs on impulsivity in both the KSP and TCI personality measures. Furthermore, when comparing PMs versus EMs on SCL-90R, all PMs were found to score below the cutoff score suggestive of distress. Then PMs versus the subgroup of EMs who scored below the cutoff point (those without distress and presenting 'positive affect') scored higher not only in impulsivity but also in perfectionism on the TCI. Impulsivity can be functional in the absence of emotional distress, but dysfunctional when used to deal with it. Perfectionism is one of the subscales of the temperament trait of persistence. It can also be functional and increase behavioral performance if not associated with pathological affective disorders, in which case it would measure the pursuit of unrealistic standards and reduce performance. These results suggested that psychologically healthy PMs versus EMs present greater impulsivity and perfectionism in the absence of distress. As previously discussed [35], these findings also suggested the relevance of controlling for psychopathology even in healthy volunteer studies because psychopathology may bias results on personality. In particular, the analysis of differences in the personality of those EMs who scored below and above the cutoff point for distress showed that these 2 groups displayed differences in all KSP scales except impulsivity and monotony avoidance.

According to present knowledge, some of the results seem to allow us to hypothesize about the main personality characteristics of CYP2D6 PMs, which seem mainly related to impulsiveness and anxiety-related traits (e.g. perfectionism). These results will be discussed in the light of the hypothesized potential hyposerotonergic/hyperdopaminergic balance in PMs. The relevance of other neurotransmitters and receptors will also be discussed.

'Impulsiveness with Anxiety/Perfectionism' and 5-HT/DA Balance

First, a wide variation in impulsive responses has long been observed in relation to 5-HT deficits [54]. In particular, it appears that 5-HT deficits influence the cognitive processing of emotional cues, with unexpected cues being changed from rewarding to punishing. Furthermore, impulsivity has also recently been shown to apparently be related to both hypo- and hyperfunctioning of the DA system [55, 56], suggesting an inverted U-shaped relationship between DA and impulsivity. Therefore, it is suggested that impulsivity encompasses a variety of related phenomena that may differ in their biological bases due to the range of behaviors that the term may describe. In particular, while impulsivity is part of normal everyday behavior, it can also be associated with neuropsychiatric disorders such as attention-deficit/hyperactivity disorder and mania which are related to hypo- and hyperdopaminergic tone, respectively [57–59]. Second, perfectionism and persistence (the temperament trait in which the TCI includes perfectionism) have been related to increased DA and decreased 5-HT synthesis. Perfectionism has recently been related to the –521C allele of the promoter region of the dopamine D4 receptor *(DRD4)* gene, which increases the efficiency of the gene expression in comparison with the –521T allele [60]. Since this gene codes for a protein distributed in brain areas relevant for the regulation/motivation of cognition, emotion and behavior, as does *CYP2D6* (frontal cortex, striatum, hippocampus, cerebellum) [11], greater extracellular dopaminergic tone in such brain areas might influence perfectionism.

On the other hand, persistence (hard-working and perfectionist individuals) was associated with the presence of homozygosity for catechol O-methyltransferase (COMT; *val/val* or *met/met* genotypes) in conjunction with the short allele of the 5-HT transporter gene *(5-HTTLPR)* [61]. Another significant interaction with persistence has been found between the dopamine transporter (DAT1) and the 102T/C polymorphism of the 5-HT_{2A} receptor gene *5HT2A* homo-/heterozygous gene variants [62]. These data suggest that an interaction between greater DA functioning and lower 5-HT in healthy volunteers might influence the temperamental personality traits of persistence and perfectionism.

CYP2D6 PMs and Personality: 'Psychopathology' in Healthy Volunteers
The aforementioned results [35] related to greater impulsivity and perfectionism in PMs have normally been associated with personality psychopathology. However, in that study these personality features appear in conjunction with low psychopathology. Recent accounts suggest that impulsivity is dysfunctional when associated with negative affect and the 'urgency' to dampen this distress through maladaptive coping [63]. This could be due to the individual's lack of tolerance and perseverance during frustration or fatigue. 'Perfectionism' in clinical populations is similarly characterized by anxiety due to excessively high performance standards and overly critical self-evaluations [64]. 'Perfectionism' in individuals with no psychopathology forms part of a healthy pursuit of excellence associated with higher academic achievement and aptitude test performance [65].

These findings appear to be consistent with the personality characteristics found in PMs compared with EMs adjusted by vulnerability to psychopathology, who showed greater KSP 'impulsiveness', but also greater perfectionism (pursuit of personal high standards despite fatigue). These results appear also to be in keeping with animal studies that induced fatigue through sustained exercise, in which fatigue correlated with increased 5-HT and decreased DA extracellular concentrations [66].

The 5-HT/DA balance hypothesis may help to make sense of the apparently conflicting results of studies analyzing the relationship between CYP2D6 and personality in healthy volunteers. For instance, lower psychasthenia implies greater energy, motivation, or functional impulsivity [29], and greater consciousness may reflect a construct similar to functional perfectionism [36]. Future research needs to address gene-adverse environment interactions by controlling the effect on personality of psychopathology or the induction of distress [48, 49, 67], and by controlling 5-HT and DA levels in PMs versus EMs. It will also need to consider whether there might be variation in PM personality across populations as a result of the influence of other genes and environmental factors.

CYP2D6 PMs and Vulnerability to Psychopathology: Studies on Patients
The findings described above on the relationship between *CYP2D6* genetic polymorphism and personality traits and psychological functioning may help to understand the involvement of CYP2D6 in psychopathology. CYP2D6 and related 5-HT and DA polymorphic genes may lead to differences in schizophrenia patients with regard to their variability in antipsychotic drug response and their constitutive psychophysiological processes and symptoms. We found a lower frequency of PMs among schizophrenia patients [68], consistent with earlier studies which also reported a lower fre-

quency of PMs in schizophrenic patients than in healthy volunteers [69, 70]. Other studies, however, found no such differences between the observed and expected frequency of CYP2D6 PMs in schizophrenic patients [71–75]. These other studies need to be interpreted with caution since they only analyzed 2 or 3 defective variant *CYP2D6* alleles (**3, *4*, and/or **10*) as causing PM status, and/or used a different control group, e.g. non-psychotics. To summarize, in the light of the present data, the association between *CYP2D6* genetic polymorphism and schizophrenia is still controversial. The reported differences in *CYP2D6* inactive alleles between patients and healthy volunteers [68] needs to be replicated.

Given the relationship between the personality characteristic of anxiety in PMs and the hypothesized lower 5-HT in these individuals, a study tested whether elderly CYP2D6 individuals carrying 2 null alleles (**4/*4*) were more predisposed to anxiety and depression disorders [76]. However, no such associations were found, although another study found an increased frequency of UMs among women with late pregnancy or post-partum depressive symptoms [77].

CYP2D6 PMs and Neurocognition

Low 5-HT or increased DA in the brain have also been related to alterations in neurocognition. In particular, low 5-HT has been associated with exaggerated aversive and impulsive cognitive processing of emotional stimuli [54]. Optimal DA function in brain has been shown to be crucial for motor activity, motivation, and such cognitive processes as attention and memory [78]. Consequently, we decided to explore whether PMs and EMs presented differences in the control of cognitive functioning under distress [35]. Participants were evaluated on objective, computerized, non-linguistic, and culturally blind cognitive tests (www.cantab.com). In all the cognitive functions assessed, PMs were only different from EMs in 'rapid visual information processing' (RVP), a test of sustained attention. When controlling for general psychopathology, PMs also showed better performance on spatial working memory, a cognitive function related to RVP which may also prevent disorders such as schizophrenia [35]. Good performance in these functions also requires optimal dopaminergic function [79], but does not appear to depend on 5-HT [80].

Acknowledgements

This study was supported by Plan Nacional de Investigación Científica, Desarrollo e Innovación Tecnológica (I+D+I) and Fondo Social Europeo of the European Union (FEDER), Instituto de Salud Carlos III-FIS Research Grant (PI06/1681) and Cibersam, and a grant to P.D. (CP06/0030), Ministerio de Educación y Ciencia (SAF2006-13589), and coordinated in the network Red Iberoamericana de Farmacogenética y Farmacogenómica (CYTED 206RT0290).

References

1 Dorado P, Berecz R, Peñas-Lledó EM, Cáceres MC, Llerena A: Clinical implications of CYP2D6 genetic polymorphism during treatment with antipsychotic drugs. Curr Drug Targets 2006;7:671–680.

2 Dorado P, Peñas-Lledó EM, Llerena A: CYP2D6 polymorphism: implications for antipsychotic drug response, schizophrenia and personality traits. Pharmacogenomics 2007;8:1597–1608.

3 Llerena A, Dorado P, Peñas-Lledó EM: Pharmacogenetics of debrisoquine and its use as a marker for CYP2D6 hydroxylation capacity. Pharmacogenomics 2009;10:17–28.

4 Kawashima H, Sequeira DJ, Nelson DR, Strobel HW: Genomic cloning and protein expression of a novel rat brain cytochrome P-450 CYP2D18* catalyzing imipramine N-demethylation. J Biol Chem 1996;271:28176–28180.

5 Wyss A, Gustafsson JA, Warner M: Cytochromes P450 of the 2D subfamily in rat brain. Mol Pharmacol 1995;47:1148–1155.

6 Hiroi T, Imaoka S, Funae Y: Dopamine formation from tyramine by CYP2D6. Biochem Biophys Res Commun 1998;249:838–843.

7 Tyndale RF, Li Y, Li NY, Messina E, Miksys S, Sellers EM: Characterization of cytochrome P-450 2D1 activity in rat brain: high-affinity kinetics for dextromethorphan. Drug Metab Dispos 1999;27:924–930.

8 Norris PJ, Hardwick JP, Emson PC: Regional distribution of cytochrome P450 2D1 in the rat central nervous system. J Comp Neurol 1996;366:244–258.

9 Riedl AG, Watts PM, Edwards RJ, Schulz-Utermoehl T, Boobis AR, Jenner P, Marsden CD: Expression and localisation of CYP2D enzymes in rat basal ganglia. Brain Res 1999;822:175–191.

10 Voirol P, Jonzier-Perey M, Porchet F, Reymond MJ, Janzer RC, Bouras C, Strobel HW, Kosel M, Eap CB, Baumann P: Cytochrome P-450 activities in human and rat brain microsomes. Brain Res 2000;855:235–243.

11 Miksys SL, Tyndale RF: Drug-metabolizing cytochrome P450s in the brain. J. Psychiatry Neurosci 2002;27:406–415.

12 Fonne-Pfister R, Bargetzi MJ, Meyer UA: MPTP, the neurotoxin inducing Parkinson's disease, is a potent competitive inhibitor of human and rat cytochrome P450 isozymes (P450bufI, P450db1) catalyzing debrisoquine 4-hydroxylation. Biochem Biophys Res Commun 1987;148:1144–1150.

13 McFadyen MC, Melvin WT, Murray GI: Cytochrome P450 in normal human brain and brain tumours. Biochem Soc Trans 1997;25:S577.

14 Gilham DE, Cairns W, Paine MJ, Modi S, Poulsom R, Roberts GC, Wolf CR: Metabolism of MPTP by cytochrome P4502D6 and the demonstration of 2D6 mRNA in human foetal and adult brain by in situ hybridization. Xenobiotica 1997;27:111–125.

15 Siegle I, Fritz P, Eckhardt K, Zanger UM, Eichelbaum M: Cellular localization and regional distribution of CYP2D6 mRNA and protein expression in human brain. Pharmacogenetics 2001;11: 237–245.

16 Llerena A, Edman G, Cobaleda J, Benítez J, Schalling D, Bertilsson L: Relationship between personality and debrisoquine hydroxylation capacity. Suggestion of an endogenous neuroactive substrate or product of the cytochrome P-4502D6. Acta Psychiatr Scand 1993;87:23–28.

17 Bromek E, Haduch A, Daniel WA: The ability of cytochrome P450 2D isoforms to synthesize dopamine in the brain: an in vitro study. Eur J Pharmacol 2010;626:171–178.

18 Yu AM, Idle JR, Krausz KW, Küpfer A, Gonzalez FJ: Contribution of individual cytochrome P450 isozymes to the O-demethylation of the psychotropic beta-carboline alkaloids harmaline and harmine. J Pharmacol Exp Ther 2003;305:315–322.

19 Yu AM, Idle JR, Byrd LG, Krausz KW, Küpfer A, Gonzalez FJ: Regeneration of serotonin from 5-methoxytryptamine by polymorphic human CYP2D6. Pharmacogenetics 2003;13:173–181.

20 Yu AM, Idle JR, Gonzalez FJ: Polymorphic cytochrome P450 2D6: humanized mouse model and endogenous substrates. Drug Metab Rev 2004;36: 243–277.

21 Kirchheiner J, Henckel HB, Franke L, Meineke I, Tzvetkov M, Uebelhack R, Roots I, Brockmöller J: Impact of the CYP2D6 ultra-rapid metabolizer genotype on doxepin pharmacokinetics and serotonin in platelets. Pharmacogenet Genomics 2005; 15:579–587.

22 Ozdemir V, Gunes A, Dahl ML, Scordo MG, Williams-Jones B, Someya T: Could endogenous substrates of drug-metabolizing enzymes influence constitutive physiology and drug target responsiveness? Pharmacogenomics 2006;7:1199–1210.

23 Ozdemir V, Bertilsson L, Miura J, Carpenter E, Reist C, Harper P, Widén J, Svensson JO, Albers LJ, Kennedy JL, Endrenyi L, Kalow W: CYP2D6 genotype in relation to perphenazine concentration and pituitary pharmacodynamic tissue sensitivity in Asians: CYP2D6-serotonin-dopamine crosstalk revisited. Pharmacogenet Genomics 2007;17:339–347.

24 Frankle WG, Lombardo I, New AS, Goodman M, Talbot PS, Huang Y, Hwang DR, Slifstein M, Curry S, Abi-Dargham A, Laruelle M, Siever LJ: Brain serotonin transporter distribution in subjects with impulsive aggressivity: a positron emission study with [^{11}C]McN 5652. Am J Psychiatry 2005;162: 915–923.
25 Carli M, Baviera M, Invernizzi RW, Balducci C: Dissociable contribution of 5-HT1A and 5-HT2A receptors in the medial prefrontal cortex to different aspects of executive control such as impulsivity and compulsive perseveration in rats. Neuropsychopharmacology 2006;31:757–767.
26 Snider NT, Sikora MJ, Sridar C, Feuerstein TJ, Rae JM, Hollenberg PF: The endocannabinoid anandamide is a substrate for the human polymorphic cytochrome P450 2D6. J Pharmacol Exp Ther 2008; 327:538–545.
27 Hiroi T, Kishimoto W, Chow T, Imaoka S, Igarashi T, Funae Y: Progesterone oxidation by cytochrome P450 2D isoforms in the brain. Endocrinology 2001;142:3901–3908.
28 Kishimoto W, Hiroi T, Shiraishi M, Osada M, Imaoka S, Kominami S, Igarashi T, Funae Y: Cytochrome P450 2D catalyze steroid 21-hydroxylation in the brain. Endocrinology 2004;145:699–705.
29 Bertilsson L, Alm C, De Las Carreras C, Widen J, Edman G, Schalling D: Debrisoquine hydroxylation polymorphism and personality. Lancet 1989; 1:555.
30 Llerena A, Cobaleda J, Benítez J: Debrisoquine hydroxylation phenotypes in healthy volunteers. Lancet 1989;8651:1398.
31 Ortet G, Torrubia R: Spanish language version of the Karolinska Scales of Personality (KSP): First data. VI Eur Conf Personality, Groningen, 1992.
32 Ortet G, Ibáñez M, Llerena A, Torrubia R: The underlying traits of the Karolinska Scales of Personality (KSP). Eur J Psychological Assess 2002;18:139–148.
33 Aklillu E, Kalow W, Endrenyi L, Harper P, Miura J, Ozdemir V: CYP2D6 and DRD2 genes differentially impact pharmacodynamic sensitivity and time course of prolactin response to perphenazine. Pharmacogenet Genomics 2007;17:989–993.
34 González I, Peñas-Lledó EM, Pérez B, Dorado P, Alvarez M, Llerena A: Relation between CYP2D6 phenotype and genotype and personality in healthy volunteers. Pharmacogenomics 2008;9: 833–840.
35 Peñas-Lledó EM, Dorado P, Pacheco R, González I, Llerena A: Relation between CYP2D6 genotype, personality, neurocognition and overall psychopathology in healthy volunteers. Pharmacogenomics 2009;10:1111–1120.
36 Kirchheiner J, Lang U, Stamm T, Sander T, Gallinat J: Association of CYP2D6 genotypes and personality traits in healthy individuals. J Clin Psychopharmacol 2006;26:440–442.
37 Suzuki E, Kitao Y, Ono Y, Iijima Y, Inada T: Cytochrome P450 2D6 polymorphism and character traits. Psychiatr Genet 2003;13:111–113.
38 Iwashima K, Yasui-Furukori N, Kaneda A, Saito M, Nakagami T, Sato Y, Kaneko S: No association between CYP2D6 polymorphisms and personality trait in Japanese. Br J Clin Pharmacol 2007;64:96–99.
39 Costa PT, McCrae RR: NEO personality Inventory professional manual. Odessa, Psychological Assessment Resources, 1992.
40 Salgado JF: The five factor model of personality and job performance in the European community. J Appl Psychol 1997;82:30–43.
41 Gan SH, Ismail R, Wan Adnan WA, Zulmi W, Kumaraswamy N, Larmie ET: Relationship between type A and B personality and debrisoquine hydroxylation capacity. Br J Clin Pharmacol 2004;57:785–789.
42 Roberts RL, Luty SE, Mulder RT, Joyce PR, Kennedy MA: Association between cytochrome P450 2D6 genotype and harm avoidance. Am J Med Genet B Neuropsychiatr Genet 2004;127:90–93.
43 Mulder RT, Joyce PR, Sullivan PF, Bulik CM, Carter FA: The relationship among three models of personality psychopathology: DSM-III-R personality disorder, TCI scores and DSQ defences. Psychol Med 1999;29:943–951.
44 Lasagna L, Von Felsinger JM: The volunteer subject in research. Science 1954;120:359–361.
45 Pollin W, Perlin S: Psychiatric evaluation of 'normal control' volunteers. Am J Psychiatry 1958;115:129–133.
46 Halbreich U, Bakhai Y, Bacon KB, Goldstein S, Asnis GM, Endicott J, Lesser J: The normalcy of self-proclaimed 'normal volunteers'. Am J Psychiatry 1989;146:1052–1055.
47 Thaker GK, Moran M, Lahti A, Adami H, Tamminga C: Psychiatric morbidity in research volunteers. Arch Gen Psychiatry 1990;47:980.
48 Shtasel DL, Gur RE, Mozley PD, Richards J, Taleff MM, Heimberg C, Gallacher F, Gur RC: Volunteers for biomedical research: recruitment and screening of normal controls. Arch Gen Psychiatry 1991;48: 1022–1025.
49 Tishler CL, Apseloff G, Bartholomae S, Reiss NS, Rhodes AR, Singh A: Are normal healthy research volunteers psychologically healthy? A pilot investigation. Exp Clin Psychopharmacol 2007;15:539–545.

50 Schechter D, Lebovitch R: Normal controls are expensive to find: methods to improve cost-effectiveness of the screening evaluation. Psychiatry Res 2005;136:69–78.
51 Derogatis LR: Misuse of the Symptom Checklist 90. Arch Gen Psychiatry 1983;40:1152–1153.
52 Gustavsson JP, Asberg M, Schalling D: The healthy control subject in psychiatric research: impulsiveness and volunteer bias. Acta Psychiatr Scand 1997;96:325–328.
53 Almeida L, Kashdan TB, Nunes T, Coelho R, Albino-Teixeira A, Soares-da-Silva P: Who volunteers for phase I clinical trials? Influences of anxiety, social anxiety and depressive symptoms on self-selection and the reporting of adverse events. Eur J Clin Pharmacol 2008;64:575–582.
54 Cools R, Roberts AC, Robbins TW: Serotoninergic regulation of emotional and behavioural control processes. Trends Cogn Sci 2008;12:31–40.
55 Williams J, Dayan P: Dopamine, learning, and impulsivity: a biological account of attention-deficit/hyperactivity disorder. J Child Adolesc Psychopharmacol 2005;15:160–179.
56 Pattij T, Vanderschuren LJ: The neuropharmacology of impulsive behaviour. Trends Pharmacol Sci 2008;29:192–199.
57 Russell VA: Dopamine hypofunction possibly results from a defect in glutamate-stimulated release of dopamine in the nucleus accumbens shell of a rat model for attention deficit hyperactivity disorder – the spontaneously hypertensive rat. Neurosci Biobehav Rev 2003;27:671–682.
58 Ralph-Williams RJ, Paulus MP, Zhuang X, Hen R, Geyer MA: Valproate attenuates hyperactive and perseverative behaviors in mutant mice with a dysregulated dopamine system. Biol Psychiatry 2003;53:352–359.
59 Abdolmaleky HM, Smith CL, Zhou JR, Thiagalingam S: Epigenetic alterations of the dopaminergic system in major psychiatric disorders. Methods Mol Biol 2008;448:187–212.
60 Bachner-Melman R, Lerer E, Zohar AH, Kremer I, Elizur Y, Nemanov L, Golan M, Blank S, Gritsenko I, Ebstein RP: Anorexia nervosa, perfectionism, and dopamine D4 receptor (DRD4). Am J Med Genet B Neuropsychiatr Genet 2007;144B:748–756.
61 Benjamin J, Osher Y, Lichtenberg P, Bachner-Melman R, Gritsenko I, Kotler M, Belmaker RH, Valsky V, Drendel M, Ebstein RP: An interaction between the catechol O-methyltransferase and serotonin transporter promoter region polymorphisms contributes to tridimensional personality questionnaire persistence scores in normal subjects. Neuropsychobiology 2000;41:48–53.
62 Schosser A, Fuchs K, Scharl T, Schloegelhofer M, Kindler J, Mossaheb N, Kaufmann RM, Leisch F, Kasper S, Sieghart W, Aschauer HN: Interaction between serotonin 5-HT2A receptor gene and dopamine transporter (DAT1) gene polymorphisms influences personality trait of persistence in Austrian Caucasians. World J Biol Psychiatry 2008;1:1–8.
63 Anestis MD, Selby EA, Joiner TE: The role of urgency in maladaptive behaviors. Behav Res Ther 2007;45:3018–3029.
64 Shafran R, Cooper Z, Fairburn CG: Clinical perfectionism: a cognitive-behavioural analysis. Behav Res Ther 2002;40:773–791.
65 Stoeber J, Kersting M: Perfectionism and aptitude test performance: testes who strive for perfection achieve better test results. Pers Individ Dif 2007;42:1093–1103.
66 Bailey SP, Davis JM, Ahlborn EN: Neuroendocrine and substrate responses to altered brain 5-HT activity during prolonged exercise to fatigue. J Appl Physiol 1993;74:3006–3012.
67 Flaa A, Ekeberg O, Kjeldsen SE, Rostrup M: Personality may influence reactivity to stress. Biopsychosoc Med 2007;1:5.
68 Llerena A, Dorado P, Peñas-Lledó EM, Cáceres MC, De la Rubia A: Low frequency of CYP2D6 poor metabolizers among schizophrenia patients. Pharmacogenomics J 2007;7:408–410.
69 Dahl AA, Løwert A, Asserson S, Bjarking L, Berglund J, Kristensen F, Norum D, Tønseth S, Bayer L, Mæhlum E: Hydroxylation polymorphism of debrisoquine hydroxylase (CYP2D6) in patients with schizophrenia in Norway and Denmark. Hum Psychopharmacol 1998;13:509–511.
70 Brockmöller J, Kirchheiner J, Schmider J, Walter S, Sachse C, Müller-Oerlinghausen B, Roots I: The impact of the CYP2D6 polymorphism on haloperidol pharmacokinetics and on the outcome of haloperidol treatment. Clin Pharmacol Ther 2002;72:438–452.
71 Dawson E, Powell JF, Nothen MM, Crocq MA, Lanczik M, Korner J, Rietschel M, van Os J, Wright P, Gill M: An association study of debrisoquine hydroxylase (CYP2D6) polymorphisms in schizophrenia. Psychiatr Genet 1994;4:215–218.
72 Daniels J, Williams J, Asherson P, McGuffin P, Owen M: No association between schizophrenia and polymorphisms within the genes for debrisoquine 4-hydroxylase (CYP2D6) and the dopamine transporter (DAT). Am J Med Genet 1995;60:85–87.
73 Pirmohamed M, Wild MJ, Kitteringham NR, O'Brien K, Buchan IE, Back DJ, Park BK: Lack association between schizophrenia and the CYP2D6 gene polymorphisms. Am J Med Genet 1996;67:236–237.

74 Jonsson EG, Dahl ML, Roh HK, Jerling M, Sedvall GC: Lack of association between debrisoquine 4-hydroxylase (CYP2D6) gene polymorphisms and schizophrenia. Psychiatr Genet 1998;8:25–28.
75 Chen CH, Hung CC, Wei FC, Koong FJ: Debrisoquine 4-hydroxylase (CYP2D6) genetic polymorphisms and susceptibility to schizophrenia in Chinese patients from Taiwan. Psychiatr Genet 2001; 11:153–155.
76 Bijl MJ, Luijendijk HJ, van den Berg JF, Visser LE, van Schaik RH, Hofman A, Vulto AG, van Gelder T, Tiemeier H, Stricker BH: Association between the CYP2D6*4 polymorphism and depression or anxiety in the elderly. Pharmacogenomics 2009;10: 541–547.
77 Josefsson A, Sydsjö G, Berg G, Dahl ML, Wadelius M, Nordin C: CYP2D6 genotypes and depressive symptoms during late pregnancy and postpartum. Nord J Psychiatry 2004;58:61–64.
78 Cools R: Role of dopamine in the motivational and cognitive control of behaviour. Neuroscientist 2008;14:381–395.
79 Bellgrove MA, Mattingley JB: Molecular genetics of attention. Ann N Y Acad Sci 2008;1129:200–212.
80 Matrenza C, Hughes JM, Kemp AH, Wesnes KA, Harrison BJ, Nathan PJ: Simultaneous depletion of serotonin and catecholamines impairs sustained attention in healthy female subjects without affecting learning and memory. J Psychopharmacol 2004;18:21–31.

Prof. Adrián LLerena
CICAB-CAIBER, Centro de Investigación Clínica Area de Salud de Badajoz
Hospital Universitario Infanta Cristina, Servicio Extremeño de Salud, Fundesalud
Avda. de Elvas s/n
ES–06080 Badajoz (Spain)
Tel./Fax +34 924 218 040, E-Mail allerena@unex.es

Schwab M, Kaschka WP, Spina E (eds): Pharmacogenomics in Psychiatry.
Adv Biol Psychiatry. Basel, Karger, 2010, vol 25, pp 46–57

Pharmacogenetics of Schizophrenia: Bringing 'Order to Chaos' in the Psychopharmacology of Schizophrenia?

Peter F. Buckley · Del D. Miller · Adriana Foster

Department of Psychiatry, Medical College of Georgia, Augusta, Ga., USA

Abstract

The inherent wide interindividual variability in response and tolerability of side effects in the treatment of schizophrenia itself provides a compelling rationale for the great potential of pharmacogenetics. Moreover, this potential is consonant with the broader public health directive toward personalized medicine. Within schizophrenia, the progress thus far has been modest in both the treatment response and adverse effect domains. This chapter chronicles the progress made in the quest for pharmacogenetic predictors in the treatment of schizophrenia.

By any measure – and especially with regard to its treatment – schizophrenia is a highly heterogeneous disorder. A continued debate in our field that is of great relevance to the area of pharmacogenetics is whether schizophrenia is actually etiologically heterogeneous – that is a collection of several conditions that arise from different pathobiological bases – or whether schizophrenia is (merely) symptomatically heterogeneous [1]. Certainly, every clinician knows well that the presentation and course of illness varies widely between patients. Irrespective of whether you ascribe this heterogeneity to neurobiology or course alone, this variability in schizophrenia itself is the 'baseline condition' upon which pharmacogenetic examinations begin. This is an important consideration.

Clinicians also know that there is wide variability in patients' response and tolerability of any given antipsychotic medication. The advent of second-generation antipsychotic medications (SGAs) alongside the first-generation antipsychotic medications (FGAs) has broadened the treatment options for patients. However, in large part the dilemma remains the same: at present, the selection of an antipsychotic is more on a 'trial and error' basis rather than based upon any robust rationale.

Table 1. Current contextual issues in the psychopharmacology of schizophrenia

Diagnostic	Drug choice	Drug profile
– Overlap between schizophrenia and bipolar disorder. – Is schizoaffective disorder a valid and useful nosological entity? – Can we really identify schizophrenia in its prodromal stages?	– Are older and new anti-psychotics really different in efficacy? – Do SGAs really differ in efficacy among each other? – What is the best drug to start with? – When should other drugs be used? – Does antipsychotic polypharmacy work?	– How do we predict and manage weight gain and metabolic disturbances? – Is dopamine D_2 receptor binding necessary and sufficient for antipsychotic efficacy? – How do you balance drug efficacy and drug tolerability over the course of illness? – Can biomarkers help to guide clinical decisions?

Set against both the variability of schizophrenia and its treatment, pharmacogenetics at least offers the promise of 'bringing order to the chaos' in the psychopharmacology of schizophrenia [2]. However, 'promise' is the operative term, and there are substantial theoretical and methodological challenges in this emergent field of pharmacogenetics. Thus, in this chapter we will describe and discuss these important contextual issues rather than simply recount findings from disparate pharmacogenetics studies. We will also, however, enumerate key studies that provide important findings for treatment responsivity and medication tolerability in schizophrenia.

Pharmacotherapy of Schizophrenia: Contextual Issues for Pharmacogenetics

In considering pharmacogenetics, the state of pharmacotherapy in schizophrenia first needs to be briefly reviewed (table 1). Several excellent recent reviews and meta-analyses provide the reader with a more comprehensive evaluation. At the present time, our field is hotly debating the relative merits of FGAs versus SGAs [3, 4]. Several large pragmatic studies have been published [5–9] which, when taken collectively, affirm the earlier observation of great inherent variability between patients so that 'mapping the right drug to the right patient' remains challenging. Thus far, drugs with antipsychotic activity appear to bear at least some relationship to a blockade of dopamine receptors. Antipsychotic efficacy appears to be tied – at least to some extent – to dopamine (D_2) blockade. Recent enthusiasm that glutamate alone might be a distinct target has been dampened by the results of a recent trial of a glutamatergic drug that had shown initial promise [10]. Additionally, these drugs have highly variable pharmacologic profiles at several other neuroreceptors. This has also been an important area of focus in pharmacogenetics (see below). Perhaps the area of greatest clinical distinction between antipsychotics is in adverse effect profiles. Extrapyramidal side effects and tardive dyskinesia (TD) are more common with FGAs than SGAs [11]. Rates of TD are still about 10 times less with SGAs. On the other hand, SGAs are more associated with weight gain and metabolic disturbances [12]. In the large CATIE study, 40% of patients met the criteria for the metabolic syndrome [13]. In a first-episode study (CAFE), 13% of patients met the criteria for the metabolic syndrome [14]. Prediction of weight gain and risk for metabolic syndrome has been a productive area of current pharmacogenetic research (see below).

Brief Overview of the Genetics of Schizophrenia: Implications for Pharmacogenetics

Although the genetics of schizophrenia are highly complex, overwhelmingly the evidence from familial, twin, and now a host of association studies collectively points to a genetic basis for schizophrenia [15–19]. While a comprehensive review of genetic studies is clearly beyond the scope and intent of this section of the paper [for a synthesis of recent findings, see 15, 17, 19], most notably the study of the genetics of schizophrenia has advanced alongside traditional familial and twin association studies to now also become increasingly molecular in focus [18, 20]. There has been a great deal of interest in polymorphisms in the Val/Met alleles of the catecholamine-O-methyl transferase gene in explaining frontal lobe functional deficits in schizophrenia [21]. Many other studies have shown abnormalities in several genes that code for neurodevelopment (e.g. dysbindin, neuroregulin, DISC, SNAP-25) and for trophic factors (e.g. BDNF) [16, 18, 22, 23]. These 'susceptibility genes' regulate proteins and/or biological processes that have also been implicated through other neurobiological (e.g. postmortem) studies in schizophrenia. For example, dysbindin is a neurodevelopmental protein gene that is found on chromosome 6. Decreased levels of dysbindin mRNA have been noted in the dorsolateral prefrontal cortex in schizophrenia patients [24], and variants in expression of components of the dysbindin gene have been reported in patients with schizophrenia [25–27]. Studies of large pedigrees have also shown a linkage signal on chromosome 10 as well as genetic variations in the gene that encodes for neuregulin [28, 29] – another neurodevelopmental gene that has been implicated in schizophrenia [30]. At present, these genetics investigations do not converge in mechanistic approaches that are used to inform pharmacogenetic investigations. Perhaps some greater confluence may occur as the field of pharmacogenetics matures further.

Findings Thus Far from Pharmacogenetic Studies in Schizophrenia

Pharmacogenetic studies of antipsychotic response and adverse effects in schizophrenia have examined both the pharmacodynamic and pharmacokinetic attributes of antipsychotic medications. Research on genetic variations of pharmacodynamic factors involved in the antipsychotic response has focused on polymorphisms of genes that code for dopamine, serotonin, histamine, muscarine, glutamate and adrenergic receptors (neurotransmitters observed to be altered in patients with schizophrenia). The pharmacokinetic studies have investigated genetic variants in enzymes known to be involved in antipsychotic metabolism.

Pharmacodynamic Factors

Numerous studies have evaluated the potential of pharmacogenetics to predict the response to antipsychotic medications [31, 32]. In a first-episode study comparing risperidone and olanzapine, Lenz et al. [33] reported that a polymorphism of the dopamine D_2 promoter gene (specifically the possession of either the –241C allele or the –141C Del allele) was associated with an enhanced treatment response during 12 weeks of treatment. There were no differences between either risperidone- or olanzapine-treated patients. Lane et al. [34] examined another polymorphism of the dopamine D_2 receptor – in this instance a polymorphism of serine (Ser 311 Cys) – in 123 patients with schizophrenia. Patients with the Ser 311 Cys allele (n = 12) showed a more robust response to antipsychotics. In a study of Chinese patients with first episode psychosis, Reynolds et al. [35] found no association between another dopamine receptor polymorphism – the Ta21A polymorphism of the dopamine D_3 receptor – and treatment response in 117 patients treated for 10 weeks with either risperidone or chlorpromazine.

However, Reynolds et al. [36] did find an association between the polymorphism (–759D C/T) in the 5HTR2C promoter region that was associated with improvements in general and negative (but not positive) symptoms. Ellingrod et al. [37] found a relationship between response to olanzapine and another 5HTR2C polymorphism (Cys 23 Ser) in a study of 41 patients with chronic schizophrenia. Polymorphisms of the 5HT2 A receptor have been shown to be associated with treatment response, most notably in an initial and influential early study by Arranz et al. [38] which reported that possession of the 102C allele of the 5HT2 A receptor predicted a poor response to clozapine. Lane et al. [39] reported contradictory findings in relation to treatment with risperidone.

Yamanouchi et al. [40] found no association between 5HT2A polymorphisms and treatment response in a short-term study of 73 patients with schizophrenia. Lane et al. [34] reported several associations between polymorphisms of the 5HTR6 gene and treatment response. Lin et al. [41] examined treatment response in relation to a polymorphism of a gene that codes for P-glycoprotein, which has been shown to transport certain SGAs across the blood brain barrier. They found that the 3435 genotype predicted positive symptom response to olanzapine treatment. Most recently, genotypic analysis was incorporated into the registration clinical trials for a novel antipsychotic, iloperidone [42]. Here a 6-marker genetic combination was associated with treatment response to iloperidone during a 4-week study. Seventy-five percent of patients with this genetic combination showed a response to iloperidone, as opposed to the response rate of 37% among the remainder of patients. Incorporating genetic analyses into antipsychotic drug development and early clinical trials programs is a substantial advance for our field. However, to date, pharmaceutical companies have been reluctant to use this approach, particularly as it has the potential to limit or 'pigeon-hole' the use of a new compound toward a subset of patients. In contrast, federally funded treatment studies proved a great opportunity to search for pharmacogenetic markers of treatment response. In this regard, the results of pharmacogenetic evaluations of treatment response in the schizophrenia CATIE study are rather salutary [43, 44]. Despite a large sample size and excellent clinical trial methodology, the assessments and analyses of polymorphism of many of the genes noted above failed to reveal any robust relationship to treatment response. A later and broad analysis of some 2,767 polymorphisms detected some weak associations, although the authors acknowledge that this may in part be due to the large number of comparisons in this analysis [44]. In balance then, examining functional polymorphisms of both dopamine and serotonin receptor genes has provided some associations with treatment response, although overall the signal appears weak and no reproducible focus emerges from these studies.

The relationship of receptor polymorphisms to adverse effects of antipsychotic medications appears to be more robust – as evident from the literature thus far. Several studies have examined polymorphisms of the dopamine D_2, D_3, and D_4 receptor substance and presence of TD. The results of these studies have been largely positive, with exceptions noted. These associations appear more robust than for 5HT2A or 5HT2C and the presence of TD. Early on, it was reported that patients of Aschenaz Jewish ethnicity had a heightened risk of developing agranulocytosis during treatment with clozapine [45]. This relationship was not confirmed in subsequent studies. However, a commercial genetic test for clozapine-induced agranulocytosis has been developed [46]. Potentially, this could predict whether a patient might be at risk of developing agranulocytosis upon exposure to clozapine. Such a test could influence the selection of patients for clozapine and/or the closer hematologic monitoring of patients who might be at risk early on during treatment with clozapine.

Although agranulocytosis was the adverse effect of greatest concern to clinicians when clozapine became available, it soon emerged that weight gain and metabolic disturbances are more worrisome side effects of this drug – and now this also appears to be the case for all SGAs [47]. Accordingly, the prediction of weight gain and metabolic disturbances during treatment with SGAs is a major focus of pharmacogenetics in schizophrenia. Reynolds et al. [48] first reported that possession of a T allele of the 5HT2 receptor was associated with weight gain during treatment with either risperidone or chlorpromazine. They found a similar relationship for clozapine therapy, also partly replicated in a study by Miller et al. [49] in a population receiving 6 months of clozapine therapy. However, studies by Basile et al. [50] and by Tsai et al. [51] did not replicate this association in clozapine-treated patients. Ellingrod et al. [52] did replicate this association in patients who were being treated with olanzapine.

Several studies have also examined the mechanistic pathways to weight gain and metabolic disturbances. Polymorphisms in the leptin gene have been associated with weight gain [53]. Jin et al. [54] provide a comprehensive review of the relationships of leptin, weight gain, and antipsychotic treatment. Ellingrod et al. [52] evaluated the relationships between methylenetetrahydrofolate reductase (MTHFR) activity and indices of the metabolic syndrome in 58 patients with schizophrenia. They observed a 4-fold increased risk of metabolic syndrome in patients with the 677T allele of MTHFR. They also found elevated insulin levels in patients with the 677T allele. Souza et al. [55] report a meta-analysis of association studies of the GNB3 gene and weight gain with antipsychotics. Overall, there appears a more consistent pattern of pharmacogenetic associations for antipsychotic-related side effects than for therapeutic response.

Pharmacokinetic Factors

Various cytochrome P450 (CYP) isoenzymes have been shown to affect the metabolism of various antipsychotics leading to interest in whether mutations in the genes that code for the enzymes predict response and adverse effects with antipsychotics. CYP2D6 is the main metabolic pathway of a number of older antipsychotics (chlorpromazine, thioridazine and haloperidol) as well as several newer antipsychotics (risperidone and aripiprazole). In a naturalistic study of haloperidol treatment, Brockmoller et al. [56] found a trend towards increased CYP2D6 activity and lower therapeutic efficacy and significantly higher ratings of parkinsonism in poor metabolizers of CYP2D6. The other studies that have investigated CYP2D6 activity have found no relationship between CYP2D6 genotype and therapeutic effects of the older antipsychotic drugs. Several studies have shown that CYP2D6 variants did not predict response to risperidone, but predicted the ratio of the parent drug to metabolite and adverse effects [57–59]. Another CYP enzyme, CYP1A2, is involved in the main metabolic pathway of clozapine and olanzapine. However, CYP1A2 polymorphisms have not been shown to significantly influence clozapine metabolism [60], but delays in response to clozapine have been observed in individuals with the ultrametabolizer phenotype [61, 62]. In addition, the combination of high inducibility CYP1A2 alleles and smoking has been found to result in reduced clozapine plasma concentrations [63]. The CYP3A4 enzyme has been shown to be involved in the metabolism of aripiprazole, quetiapine, risperidone and to a lesser extent clozapine and ziprasidone. To date, no significant reports of an association of the identified variants of CYP3A4 with antipsychotic variability or response have been published. Likewise no signifi-

Table 2. Considerations that could impact the therapeutic potential of pharmacogenetics in schizophrenia

Patient variables	Measurement variables	Treatment variables
– Diagnosis – Illness onset and duration – Clinical heterogeneity – Medical and psychiatric comorbidities – Extent of prior treatment-refractoriness	– Definition(s) of treatment responses – Measurement scale(s) used – Duration of observation – Physiological indices (e.g. insulin sensitivity, leptin levels) and their measurement issues – Technical aspects of pharmacogenetic blood tests	– Study durations – Drug use – Presence of concomitant medications

cant response associations have been reported with the polymorphic CYP3A5, an enzyme reported to contribute to antipsychotic metabolism [64].

Thus at the present time, the use of pharmacogenetics of antipsychotic kinetics may be clinically useful for predicting dose in special cases and for certain antipsychotics, while their usefulness in predicting clinical response must be further explored.

There is also a series of studies examining the association between TD and the cytochrome P450 genes CYP2D6 and CYP1A2. The majority of studies found that mutations resulting in reduced 2D6 activity (and presumably in higher plasma concentrations of antipsychotic medications) were positively correlated with higher AIMS scores and the development of TD [65–68]. Conversely, Sachse et al. [67] found that the CYP2D6 polymorphisms did not predict TD but that CYP1A2 polymorphisms were significantly associated with TD. Basile et al. [69] reported that patients who were homozygous for the C allele of the CYP1A2 gene had significantly higher AIMS scale scores. This finding was not replicated in a study by Schulze et al. [70].

Methodological Considerations in Optimizing Pharmacogenetic 'Signals' in Schizophrenia Research

The extent to which the 'promise' of pharmacogenetics might be realized in schizophrenia research – 'bringing order to chaos' and heralding the clinical expression of personalized medicine – is in part dependent on several factors (table 2). Just as in 'classical genetics', the ability to detect any meaningful signal in pharmacogenetics studies is dependent upon careful diagnostic assessment. This may seem too intuitive to merit attention, yet the trend in clinical trials research today toward large pragmatic trials may be contributory. These studies include heterogeneous patient populations with psychiatric and medical comorbidities which might obfuscate pharmacogenetic associations, especially if these are weak and/or only observed in combination. The lack of robust associations in the CATIE schizophrenia study is noteworthy and this study, while exceptionally large, did enroll an impressively heterogeneous patient population.

Another very important consideration here is ethnicity of the study population [31, 71]. It is likely that the results in various studies of response and adverse effects may also be related to ethnicity. This is most evident for investigations of the CYP2D6 cytochrome system which is well known to vary substantially in expression according to ethnicity.

It is also unclear whether pharmacogenetic associations are likely to emerge as class related (i.e. weight gain associated with all SGAs) or drug-specific (e.g. 5HT2 receptor –759C/T association with clozapine-related weight gain but not with other SGAs). Clearly, clues in either direction would support a more tailored research focus in later studies.

It is also important to appreciate that pharmacogenetic studies are subject to the same methodological issues that conspire against positive therapeutic findings in psychopharmacologic research: Was the right dose of the antipsychotic used for the study? Was the duration of the study long enough to observe a therapeutic response? Was this a study population 'capable' of showing a therapeutic response or did the investigators unwittingly choose a more refractory patient sample for this treatment study?

It is also evident that definitions of treatment response differ substantially across studies, reflecting the current state of methodology within psychopharmacology. For example, the study of Arranz et al. [38], which showed an association between symptom response to clozapine therapy and 5HT2C, was based upon scores on the Global Assessment of Functioning Scale. This is a reasonable approach. However, of course it should not be a surprise that subsequent studies using the Brief Psychiatric Rating Scale or other measures did not replicate this initial finding. The conventions for defining treatment response in schizophrenia also vary across studies – some apply a percent change from baseline ratings, some apply composite measures, and increasingly studies are considering remission and recovery as therapeutic outcomes [72]. In terms of evaluating the capacity of pharmacogenetics to provide a predictive signal, this variance in measurement and definition of treatment response is indeed shifting ground which must surely contribute to the inconsistency in results across pharmacogenetic studies. In this regard, it may well be that adverse effect profiles may be a more tangible measure to advance pharmacogenetic investigations in schizophrenia.

Pharmacogenetics: 'Bringing Order to Chaos?'

Personalized medicine, while not quite inculcated into current clinical care, is looming large as the next transformation of healthcare [73–75]. Synderman and Dinan [76] articulate a fundamental shift from a 'find it, fix it' model of care to a 'personalize it, predict it' model (table 3). This is extremely exciting and provocative. Already, we can

Table 3. Evolution toward personalized medicine

Current approaches	Personalized care
– Disease diagnosis	– Risk identification and analysis
– Identifying overall risk factors	– Delineating personalized 'risk factor' profile
– Developing a treatment plan	– Constructing personalized care plan
– Algorithm-based care	– Personalized individual risk-based care
– Global preventative approaches	– Individual wellness and risk reduction management

Derived from Snyderman and Dinan [76].

see glimpses of this potential, as evidenced in the pharmacogenetics of anticoagulant therapy [77] and in increasingly refined and genetics-guided approaches to cancer chemotherapy [78]. Psychiatry – and in this instance, the treatment of schizophrenia – deserves no less.

If either therapeutic response and/or tolerability to antipsychotic medications have a neurobiological basis that is genetically regulated, then the promise of pharmacogenetics remains considerable. If we could select the initial choice of antipsychotic for a first-episode psychosis patient based upon his/her genetic profile, this would be paradigm-breaking for psychopharmacology. If we could predict which patient is going to develop antipsychotic-induced diabetes mellitus for a given drug, we would avoid exposure to that agent. If we knew which genes were important to treatment response with one drug and that they differed between drugs, we could make rational decisions about which drug to try next when the patient fails on the present antipsychotic. All, or even any, of these advances would represent substantial progress in psychopharmacology and they would take us well beyond the repeated 'trial and error' of current treatment.

References

1 Buckley PF, Jones PJ: Schizophrenia. A Postgraduate Textbook of Psychiatry. London, Hodder, 2009.
2 Arranz MJ, Kapur S: Pharmacogenetics in psychiatry: are we ready for widespread clinical use? Schizophr Bull 2008;34:1130–1144.
3 Tandon R, Fleischhacker WW: Comparative efficacy of antipsychotics in the treatment of schizophrenia: a critical assessment. Schizophr Res 2005;79:145–155.
4 Tandon R, Nasrallah HA, Keshavan MS: Schizophrenia, 'just the facts' 4. Clinical features and conceptualization. Schizophr Res 2009;110:1–23.
5 Lieberman JA, Stroup TS, McEvoy JP, et al: Effectiveness of antipsychotic drugs in patients with chronic schizophrenia. N Engl J Med 2005;353:1209–1223.
6 Stroup TS, Lieberman JA, McEvoy JP, et al: Effectiveness of olanzapine, quetiapine, risperidone, and ziprasidone in patients with chronic schizophrenia following discontinuation of a previous atypical antipsychotic. Am J Psychiatry 2006;163:611–622.
7 McEvoy JP, Lieberman JA, Stroup TS, et al: Effectiveness of clozapine versus olanzapine, quetiapine, and risperidone in patients with chronic schizophrenia who did not respond to prior atypical antipsychotic treatment. Am J Psychiatry 2006;163:563–565.

8 Jones PB, Barnes TR, Davies L, et al: Randomized controlled trial of the effect on quality of life of second- versus first-generation antipsychotic drugs in schizophrenia: Cost Utility of the Latest Antipsychotic Drugs in Schizophrenia Study (CUtLASS 1). Arch Gen Psychiatry 2006;63:1079–1087.

9 Kahn RS, Fleischhacker WW, Boter H: Effectiveness of antipsychotic drugs in first-episode schizophrenia and schizophreniform disorder: an open randomized clinical trial. Lancet 2008;371:1085–1097.

10 Kinon BJ, Chen L, Ascher-Svanum H: Early response to antipsychotic drug therapy as a clinical marker of subsequent response in the treatment of schizophrenia. Neuropsychopharmacology 2009; 35:581–590.

11 Miller DD, McEvoy JP, Davis SM, et al: Clinical correlates of tardive dyskinesia in schizophrenia: baseline data from the CATIE schizophrenia trial. Schizophr Res 2005;80:33–43.

12 Buckley PF, Foster A, Miller B: Schizophrenia host vulnerability and risk for metabolic disturbances during treatment with antipsychotics. Focus 2008; 4:1722–1729.

13 Nasrallah HA: Metabolic findings from the CATIE trial and their relation to tolerability. CNS Spectr 2006;11:32–39.

14 Patel JK, Buckley PF, Woolson S, Hamer RM, McEvoy JP, Perkins DO, Lieberman JA, Cafe Investigators: Metabolic profiles of second-generation antipsychotics in early psychosis: findings from the CAFE study. Schizophr Res 2009;111:9–16.

15 Delisi LE: Searching for the true genetic vulnerability for schizophrenia. Genome Med 2009;23:14.

16 Keshavan MS, Tandon R, Boutros NN, Nasrallah HA: Schizophrenia, 'just the facts': what we know in 2008. Part 3: neurobiology. Schizophr Res 2008; 106:89–107.

17 Smith M: The year in human and medical genetics: highlights of 2007–2008. Ann NY Acad Sci 2009; 1151:1–21.

18 Tan HY, Nicodemus KK, Chen Q: Genetic variation in AKT1 is linked to dopamine-associated prefrontal cortical structure and function in humans. J Clin Invest 2008;118:2200–2208.

19 Ng RM, Pearson V, Lam M, et al: What does recovery from schizophrenia mean? Perceptions of long-term patients. Int J Soc Psychiatry 2008;54:118–130.

20 Bertolino A, Fazio L, Caforio G: Functional variants of the dopamine receptor D_2 gene modulate prefronto-striatal phenotypes in schizophrenia. Brain 2009;132:417–425.

21 Goldman D, Weinberger DR, Malhotra AK: The role of COMT Val158Met in cognition. Biol Psychiatry 2009;65:e1–e2.

22 Harrison PJ: Schizophrenia susceptibility genes and their neurodevelopmental implications: focus on neuregulin 1. Novartis Found Symp 2007;288: 246–255.

23 Harrison PJ, Weinberger DR: Schizophrenia genes, gene expression, and neuropathology: on the matter of their convergence. Mol Psychiatry 2008;10: 40–68.

24 Weickert CS, Straub RE, McClintock BW, et al: Human dysbindin (DTNBP 1) gene expression in normal brain and in schizophrenic prefrontal cortex and midbrain. Arch Gen Psychiatry 2004;61:544–555.

25 Pae CU, Drago A, Kim JJ, et al: DTNBP1 haplotype influences baseline assessment scores in schizophrenic patients. Neurosci Lett 2008;440:150–154.

26 Morris DW, Murphy K, Kenny N, Purcell SM, et al: Dysbindin (DTNBP1) and the biogenesis of lysosome-related organelles complex 1 (BLOC-1): main and epistatic gene effects are potential contributors to schizophrenia susceptibility. Biol Psychiatry 2008;63:24–31.

27 Fanous AH, Van Den Oord EJ, Riley BP, et al: Relationship between a high-risk halotype in the DTNBP1 (dysbindin) gene and clinical features of schizophrenia. Am J Psychiatry 2005;162:1824–1832.

28 Chen PL, Avramopoulos D, Lasseter VK: Fine mapping on chromosome 10q22–q23 implicates neuregulin 3 in schizophrenia. Am J Hum Genet 2009; 84:21–34.

29 Wang XD, Su YA, Guo CM: Chronic antipsychotic drug administration alters the expression of neuregulin 1beta, ErbB2, ErbB3, and ErbB4 in the rat prefrontal cortex and hippocampus. Int J Neuropsychopharmacol 2008;11:553–561.

30 Mei L, Xiong WC: Neuregulin 1 in neural development, synaptic plasticity and schizophrenia. Nat Rev Neurosci 2008;9:437–452.

31 Foster A, Miller DD, Buckley PF: Pharmacogenetics and schizophrenia. Psychiatr Clin North Am 2007; 30:417–435.

32 Arranz MJ, de Leon J: Pharmacogenetics and pharmacogenomics of schizophrenia: a review of last decade of research. Mol Psychiatry 2007;12:707–747.

33 Lencz T, Robinson D, Xu K, et al: DRD2 promoter region as a predictor of sustained response to antipsychotic medication in first episode schizophrenia patients. Am J Psychiatry 2006;163:529–531.

34 Lane HY, Lee CC, Change YC, et al: Effects of dopamine D2 receptor Ser311Cys polymorphism and clinical factors on risperidone efficacy for positive and negative symptoms and social function. Int J Neuropsychopharmacol 2004;7:461–470.

35 Reynolds GP, Templeman LA, Zhang ZJ: The role of 5HT2C receptor polymorphisms in the pharmacogenetics of antipsychotic drug treatment. Prog Neuropsychopharmacol Biol Psychiatry 2005;29: 1021–1028.
36 Reynolds GP, Yao Z, Zhang X, et al: Pharmacogenetics of treatment in first-episode schizophrenia: D3 and 5-HT2C receptor polymorphism separately associate with positive and negative symptom response. Eur Neuropsychopharmacol 2005;15:143–151.
37 Ellingrod VL, Lund BC, Miller D, et al: 5-HT2A receptor promoter polymorphism, –1438G/A and negative symptom response to olanzapine in schizophrenia. Psychopharmacol Bull 2003;37:109–112.
38 Arranz M, Collier D, Sodhi M, et al: Association between clozapine response and allelic variation in 5-HT2A receptor gene. Am J Psychiatry 2002;159: 1593–1595.
39 Lane HY, Lin CC, Huany CH, et al: Risperidone response and 5-HT6 receptor gene variance: genetic association analysis with adjustment for nongenetic confounds. Schizophr Res 2004;67:63–70.
40 Yamanouchi Y, Iwata N, Suzuki T, et al: Effect of DRD2, 5-HT2A, and COMT genes on antipsychotic response to risperidone. Pharmacogenomics J 2003;3:356–361.
41 Lin YC, Ellingrod VL, Bishop JR, et al: The relationship between glycoprotein (PGP) polymorphsims and response to olanzapine treatment in schizophrenia. Ther Drug Monit 2006;28:668–672.
42 Volpi S, Potkin SG, Malhotra AK, et al: Applicability of a genetic signature for enhanced Iloperidone efficacy in the treatment of schizophrenia. J Clin Psychiatry 2009;70:801–809.
43 Grossman I, Sullivan PF, Walley N, et al: Genetic determinants of variable metabolism have impact on the clinical use of leading antipsychotics in the CATIE study. Genet Med 2008;10:720–729.
44 Need AC, Keefe RS, Ge D, Grossman I, et al: Pharmacogenetics of antipsychotic response in the CATIE trial: a candidate gene analysis. Eur J Hum Genet 2009;17:946–957.
45 Alvir JM, Lieberman JA, Safferman AZ, et al: Clozapine-induced agranulocytosis: incidence and risk factors in the United States. N Engl J Med 1993; 329:162–167.
46 Opgen-Rhein C, Dettling M: Clozapine-induced agranulocytosis and its genetic determinants. Pharmacogenomics 2008;9:1101–1111.
47 Buckley PF: Second-generation antipsychotic medications in the treatment of mood disorders: focus on aripiprazole. Drugs Today (Barc) 2005;41:5–11.
48 Reynolds GP, Zhang ZI, Zhang XB: Association of antipsychotic induced weight gain with a 5HT2 receptor gene polymorphism. Lancet 2002;359:2086–2087.
49 Miller DD, Ellingrod VL, Holman TL, et al: Clozapine-induced weight gain associated with the 5HT2C receptor –759C/T polymorphism. Am J Med Genet B Neuropsychiatr Genet 2005;133:97–100.
50 Basile VS, Ozemir V, Masellis M, et al: Lack of association between serotonin-2A receptor gen (HTR2A) polymorphisms and tardive dyskinesia in schizophrenia. Mol Psychiatry 2001;6:230–234.
51 Tsai SJ, Yu YW, Lin CH, et al: Association study of adrenergic beta3 receptor (Trp64Arg) and G-protein beta3 subunit gene (C825T) polymorphisms and weight change during clozapine treatment. Neuropsychobiology 2004;50:37–40.
52 Ellingrod VL, Bishop JR, Moline J, et al: Leptin and leptin gene polymorphisms and increases in body mass index (BMI) from olanzapine treatment in persons with schizophrenia. Curr Psychiatry 2007; 9:313–318.
53 Jin H, Meyer JM, Mudaliar S: Impact of atypical antipsychotic therapy on leptin, ghrelin, and adiponectin. Schizophr Res 2008;100:70–85.
54 Ellingrod VL, Miller DD, Taylor SF, et al: Metabolic syndrome and insulin resistance in schizophrenia patients receiving antipsychotics genotyped for the methylenetetrahydrofolate reductase (MTHFR) 677C/T and 1298A/C variants. Schizophr Res 2008;98:47–54.
55 Souza RP, DeLuca V, Muscettola G, et al: Association of antipsychotic induced weight gain and body mass index with GNB3 gene: a meta-analysis. Prog Neuropsychopharmacol Biol Psychiatry 2008;12: 32;1848–1853.
56 Brockmoller J, Kirchheiner J, Schmider J, Walter S, Sachse C, Muller-Oerlinghausen B, et al: The impact of the CYP2D6 polymorphism on haloperidol pharmacokinetics and on the outcome of haloperidol treatment. Clin Pharmacol Ther 2002;72:438–452.
57 Riedel M, Schwarz MJ, Strassnig M, Spellmann I, Muller-Arends A, Weber K, et al: Risperidone plasma levels, clinical response and side-effects. Eur Arch Psychiatry Clin Neurosci 2005;255:261–268.
58 de Leon J, Susce MT, Pan RM, Fairchild M, Koch WH, Wedlund PJ: The CYP2D6 poor metabolizer phenotype may be associated with risperidone adverse drug reactions and discontinuation. J Clin Psychiatry 2005;66:15–27.
59 Kakihara S, Yoshimura R, Shinkai K, Matsumoto C, Goto M, Kaji K, et al: Prediction of response to risperidone treatment with respect to plasma concentrations of risperidone, catecholamine metabolites, and polymorphism of cytochrome P450 2D6. Int Clin Psychopharmacol 2005;20:71–78.
60 Kootstra-Ros JE, Smallegoor W, van der Welde J: The cytochrome P450 CYP1A2 genetic polymorphisms *1F and *1D do not affect clozapine clearance in a group of schizophrenic patients. Ann Clin Biochem 2005;42:216–219.

61 Ozdemir V, Kalow W, Okey AB, Lam MS, Albers LJ, Reist C, et al: Treatment-resistance to clozapine in association with ultrarapid CYP1A2 activity and the C- >A polymorphism in intron 1 of the CYP1A2 gene: effect of grapefruit juice and low-dose fluvoxamine. J Clin Psychopharmacol 2001; 21:603–607.
62 Eap CB, Bender S, Jaquenoud SE, Cucchia G, Jonzier-Perey M, Baumann P, et al: Nonresponse to clozapine and ultrarapid CYP1A2 activity: clinical data and analysis of CYP1A2 gene. J Clin Psychopharmacol 2004;24:214–219.
63 Bondolfi G, Morel F, Crettol S, Rachid F, Baumann P, Eap CB: Increased clozapine plasma concentrations and side effects induced by smoking cessation in 2 CYP1A2 genotyped patients. Ther Drug Monit 2005;27:539–543.
64 Kalgutkar AS, Taylor TJ, Venkatakrishnan K, Isin EM: Assessment of the contributions of CYP3A4 and CYP3A5 in the metabolism of the antipsychotic agent haloperidol to its potentially neurotoxic pyridinium metabolite and effect of antidepressants on the bioactivation pathway. Drug Metab Dispos 2003;31:243–249.
65 Kapitany T, Meszaros K, Lenzinger E, et al: Genetic polymorphism for drug metabolism (CYP2D6) and tardive dyskinesia in schizophrenia. Schizophr Res 1998;32:101–106.
66 Ohmori O, Suzuki T, Kojima H, et al: Tardive dyskinesia and debrisoquine 4-hydroxylase (CYP2D6) genotype in Japanese schizophrenics. Schizophr Res 1998;32:107–113.
67 Ellingrod VL, Perry PJ, Lund BC, et al: 5HT2A and 5HT2C receptor polymorphisms and predicting clinical response to olanzapine in schizophrenia. J Clin Psychopharmacol 2002;22:622–624.
68 Sachse C, Brockmoller J, Baue S, et al: Cytochrome P450 2D6 variants in a Caucasian population: allele frequencies and phenotypic consequences. Am J Hum Genet 1997;60:284–295.
69 Basile VS, Ozdemir V, Masellis M, et al: A functional polymorphism of the cytochrome P450 1A2 (CYP1A2) gene: association with tardive dyskinesia in schizophrenia. Mol Psychiatry 2000;5:410–417.
70 Schulze TG, Schumacher J, Muller DJ, et al: Lack of association between a functional polymorphism of the cytochrome P450 1A2 (CYP1A2) gene and tardive dyskinesia in schizophrenia. Am J Med Genet 2001;105:498–501.
71 Rodriguez-Antona C, et al: CYP2D6 genotyping for psychiatric patients treated with risperidone: consideration for cost effectiveness studies. Pharmacogenomics 2009;10:685–699.
72 Kane JM: Relapse prevention in patients with schizophrenia. J Clin Psychiatry 2008;69:e11.
73 Fackler JL, McGuire AL: Paving the way to personalized genomic medicine: steps to successful implementation. Curr Pharmacogenomics Person Med 2009;7:125–132.
74 Leeder JS, Spielberg SP: Personalized medicine: reality and reality checks. Ann Pharmacother 2009; 43:963–966.
75 Guttmacher AE, McGuire AL, Ponder B, Stefansson K: Personalized genomic information: preparing for the future of genetic medicine. Nat Rev Genet 2010;11:161–165.
76 Snyderman R, Dinan MA: Improving health by taking it personally. JAMA 2010;303:363–364.
77 Kurnik D, Loebstein R, Halkin H, Gak E, Almog S: 10 years of anticoagulant pharmacogenomics: what difference will it make? A critical appraisal. Pharmacogenomics 2009;10:1955–1965.
78 Schroth W, Goetz MP, Hamann U, et al: Association between CYP2D6 polymorphisms and outcomes among women with early stage breast cancer treated with tamoxifen. JAMA 2009;302:1429–1436.

Peter F. Buckley, MD
Department of Psychiatry, Medical College of Georgia
997 St. Sebastian Way
Augusta, GA 30912 (USA)
E-Mail pbuckley@mcg.edu

Schwab M, Kaschka WP, Spina E (eds): Pharmacogenomics in Psychiatry.
Adv Biol Psychiatry. Basel, Karger, 2010, vol 25, pp 58–74

Pharmacogenomics of Depression

Alessandro Serretti[a] · Antonio Drago[a] · Edoardo Spina[b]

[a]Department of Psychiatry, University of Bologna, Bologna, and [b]Department of Clinical and Experimental Medicine and Pharmacology, University of Messina, and IRCCS Centro Neurolesi 'Bonino-Pulejo', Messina, Italy

Abstract

Pharmacotherapy of depression is characterized by poor predictability of individual response. In addition to pathophysiological and environmental factors, genetic factors appear to play an important role in determining differences in treatment outcome of antidepressant drugs (ADs). In recent years, a number of pharmacogenetic studies have been conducted on ADs, and genetic variations at the level of drug-metabolizing enzymes, drug transporters and drug targets that may influence the clinical response have been identified. Hopefully, pharmacogenetics will provide the basis for individualized pharmacotherapy of depressive disorders in order to maximize the probability of a favorable response and to minimize the risk of adverse drug reactions. In this chapter, the major findings related to the pharmacogenetics of genes involved in the pharmacokinetics and pharmacodynamics of ADs are critically reviewed.

Depression is a major psychiatric disorder, predicted to be the second leading cause of death and disability by the year 2020, which requires long-term, often life-long, pharmacological treatment [1]. Drugs currently available for the treatment of depressive disorders include older compounds, such as tricyclic antidepressants (TCAs) and monoamine oxidase inhibitors, and newer agents, such as selective serotonin reuptake inhibitors (SSRIs), noradrenaline reuptake inhibitors, serotonin-noradrenaline reuptake inhibitors, noradrenaline-dopamine reuptake inhibitors, and noradrenergic and specific serotonergic antidepressants [2]. All antidepressant drugs (ADs) target the monoaminergic systems (serotonin, noradrenaline and/or dopamine) and interfere with the metabolism, release, binding and reuptake of monoamines.

Even though ADs have successfully been used to treat depressive disorders, there is still substantial need for improvement. Response to AD therapy is often incomplete with approximately 30–40% not responding at all to the first AD given and about

60–70% not achieving remission [3]. Moreover, AD pharmacotherapy is still hampered by a delayed time of onset of clinical improvement and a variety of adverse effects. Such shortcomings of AD medication not only lead to personal suffering in both individuals and their families, but also impose considerable costs on society. Therefore, in order to reduce the patients' disability and minimize costs, it would be desirable to know in advance whether a drug is likely to be effective and tolerable. Unfortunately, at present, there is no reliable way to predict the individual's response to a specific AD before initiation of treatment and clinical and anamnestic variants were not found to be helpful for this purpose [4].

In recent years, the development of pharmacogenetics has provided more opportunities for individualized pharmacotherapy of depressive disorders [5, 6]. It is well known that the large interpatient variability in clinical response to ADs is influenced by a variety of genetic as well as pathophysiological and environmental factors. So far, pharmacogenetic studies have investigated genes involved in the pharmacokinetics and pharmacodynamics of ADs. Genetic variations at level of drug-metabolizing enzymes, drug transporters, drug targets and other biomarker genes, possibly influencing clinical response, have been identified [5–7].

In this chapter, we summarize the major findings related to the pharmacogenetics of genes affecting response to ADs.

Pharmacokinetics

Genetic variations at level of drug-metabolizing enzymes and drug transporters may affect the pharmacokinetics of ADs.

Drug-Metabolizing Enzymes

Like most psychotropic drugs, ADs are highly lipophilic agents subject to extensive biotransformation in the liver. In general, their metabolism includes initial phase I oxidative reactions, catalyzed by cytochrome P450 (CYP) enzymes, followed by phase II glucuronide conjugation.

The human CYP system consists of a superfamily of more than 50 heme-containing enzymes, located in the membranes of the smooth endoplasmic reticulum in the liver and in many extrahepatic tissues, that are responsible for the phase I oxidative reactions of many drugs and endogenous substances [8]. The CYP isoenzymes playing a major role in the biotransformation of therapeutic agents are CYP1A2, CYP2C9, CYP2C19, CYP2D6 and CYP3A4. Minor but clinically relevant isoforms include CYP2A6, CYP2B6, CYP2C8 and CYP2E1. The major enzymes involved in the biotransformation of different ADs are indicated in table 1 [9]. The metabolic activity of CYPs is genetically determined and mutations or polymorphisms in genes coding for CYP isoforms can result in enzyme variants with higher, lower or no activity, or occasionally the total absence of the enzyme. Among the

Table 1. Enzymes involved in the biotransformation of ADs (based on Spina et al. [9])

AD	Enzymes involved in biotransformation
Tricyclic antidepressants (demethylation)	CYP2C19, CYP1A2, CYP3A4
Tricyclic antidepressants (hydroxylation)	CYP2D6
Fluoxetine	CYP2D6, CYP2C9, CYP2C19, CYP3A4
Paroxetine	CYP2D6, CYP3A4
Fluvoxamine	CYP1A2, CYP2D6
Sertraline	CYP2C9, CYP2C19, CYP2D6, CYP3A4
Citalopram	CYP2C19, CYP2D6, CYP3A4
Escitalopram	CYP2C19, CYP2D6, CYP3A4
Venlafaxine	CYP2D6, CYP3A4
Duloxetine	CYP2D6, CYP1A2
Mirtazapine	CYP2D6, CYP1A2, CYP3A4
Reboxetine	CYP3A4
Bupropion	CYP2B6
Nefazodone	CYP3A4

CYP genetic polymorphisms, CYP2D6 and CP2C19 play a relevant role in the biotransformation of ADs.

Although it accounts for less than 5% of the total hepatic CYP content, CYP2D6 plays an important role in drug metabolism, being partially or entirely responsible for the oxidative biotransformation of many therapeutic agents [8]. The gene encoding for CYP2D6 is located in position 22q13.1, and it spans 4,382 bases. To date, more than 70 allelic variants have been described, some encoding an inactive or no enzyme at all, while others consist of a gene duplication. Alleles with duplication or multiduplication of a functional CYP2D6*2 gene are associated with increased CYP2D6 activity: the frequency of this condition varies from 1–2% in Swedes to up to 7–10% in Spaniards and Southern Italians. According to the inherited CYP2D6 alleles, individuals are classified as poor, intermediate, extensive or ultrarapid metabolizers. In vivo and in vitro studies have documented that many ADs – including TCAs (hydroxylation reactions), fluoxetine, paroxetine, fluvoxamine, venlafaxine and mirtazapine – are metabolized, at least in part, by CYP2D6 [7, 10]. On the other hand, the CYP2D6 polymorphism seems to have no major influence on the biotransformation of sertraline, citalopram, escitalopram, duloxetine, reboxetine and bupropion. A number of studies have investigated the relationship between the CYP2D6 polymorphism, plasma levels and clinical response to ADs that are substrates of CYP2D6 [7, 10]. There is substantial evidence indicating that genetically determined differences in pharmacokinetics may impact the outcome and risk of adverse drug reactions of ADs, with anecdotal reports describing an association between CYP2D6 defective variants and AD drug toxicity [7, 10, 11]. However, some controversy still exists concerning the influence of the CYP2D6 polymorphism on clinical response to ADs, and further investigation in large studies is required [12]. Specific dose recommendations based on CYP2D6 genotypes have even been suggested for some ADs [7, 13, 14]. Based on the available data, knowledge of the CYP2D6 metabolizer status might be useful in individualizing dose escalation schemes only for ADs with a narrow therapeutic index, such as TCAs. On the other hand, metabolizer-status-dependent dose adjustments are presumably not necessary for SSRIs and other newer ADs with a wide therapeutic window and no evidence of a clear-cut correlation between plasma levels and clinical response.

The CYP2C19 gene is found in position 10q24.1–q24.3, and it spans 90,636 bases. CYP2C19 also exhibits a clinically important genetic polymorphism [8]. The most frequent defective alleles resulting in a nonfunctional enzyme and responsible for the poor metabolizer phenotype are CYP2C19*2, the most common among Caucasians and Orientals, and CYP2C19*3, found at a frequency of about 12% among Orientals, but almost absent among Caucasians. CYP2C19 is also involved in the metabolism of some ADs. In particular, it is the major enzyme responsible for the demethylation of amitriptyline, imipramine and clomipramine, and contributes to the biotransformation of citalopram, escitalopram and sertraline. As compared to CYP2D6, CYP2C19 polymorphism appears to have a lower impact on pharmacokinetics and clinical response to ADs, so the usefulness of CYP2C19 genotyping procedures as a guide for individualization of AD dose is obviously limited [7, 13, 14].

In conclusion, the pharmacokinetics of many TCAs, some SSRIs and other ADs are significantly altered by CYP2D6 and CYP2C19 polymorphisms; however, it is still controversial whether therapeutic efficacy may be improved and/or adverse effects can be prevented by the use of genotyping procedures. Appropriate observational, longitudinal studies are needed to assess more precisely the impact of CYP polymorphisms on clinical response to ADs. With regard to this, the recent approval by the FDA of a pharmacogenetic test, the AmpliCyp CYP450 Test (Roche Molecular Systems), which assesses both polymorphic genes CYP2D6 and CYP2C19, may be of help in validating studies on personalized therapy of depression. On the other hand, if polymorphic oxidation appears to play a minor role in the clinical response to new ADs, the ability of some of these agents to act as inhibitors of CYP enzymes is of great importance [9]. New ADs differ considerably in their potential for metabolic drug interactions. Fluoxetine and paroxetine are potent inhibitors of CYP2D6, fluvoxamine markedly inhibits CYP1A2 and CYP2C19, while nefazodone is a substantial inhibitor of CYP3A4. Therefore, clinically relevant interactions are expected when these agents are coadministered with substrates of these CYP isoenzymes, especially those with a narrow therapeutic index. Duloxetine and bupropion are moderate inhibitors of CYP2D6, while sertraline will only cause significant inhibition of this isoform at high doses. Other second-generation ADs – including citalopram, escitalopram, venlafaxine, mirtazapine and reboxetine – are weak or negligible inhibitors of the different CYP isoforms and are less likely to interact with other medications.

Drug Transporters – P-Glycoprotein

P-glycoprotein is an ATP-binding transporter protein encoded by the ABCB1 gene (alternate name MDR1) on chromosome 7q21 [15]. It is a multidrug efflux transporter highly expressed in the intestine, brain, liver and kidney, which acts as a natural defense mechanism against several substrates by limiting their absorption from the gut and penetration to the brain and promoting their elimination in the bile and urine. Because of its location at the blood brain barrier, P-glycoprotein may regulate the concentration of ADs in the brain. In vitro studies and experiments in knock-out mice devoid of functional P-glycoprotein have documented that the SSRIs citalopram, sertraline and paroxetine, the TCAs trimipramine, amitriptyline, nortriptyline and doxepine and the serotonin-noradrenaline reuptake inhibitors venlafaxine are substrates of P-glycoprotein, while this may not be true for fluoxetine, bupropion and mirtazapine [16]. Several polymorphisms in ABCB1 have been identified and 3 of them, 2 synonymous SNPs (C3435T and C1236) and a non-synonymous SNP (G2677T), have been associated with altered P-glycoprotein activity. Studies investigating the influence of these functional polymorphisms on AD plasma levels, side effect profile and clinical response have given contradictory results [17–19]. In a recent study [20], common intronic polymorphisms in ABCB1 were associated with treatment outcome in patients treated with ADs substrates of P-glycoprotein (citalopram, venlafaxine and paroxetine) but not in patients receiving mirtazapine, a drug that does not appear to be a substrate of this transporter.

Pharmacodynamics

An increasing number of different AD molecules have been synthesized in the last decades, whilst the details of the pharmacodynamic events that hold the AD potential lagged behind. The lack of this knowledge is partially responsible for the incomplete efficacy of AD pharmacotreatments [3]. Pharmacogenetics holds the potential for a change: the identification of genes whose variations regulate the AD response may lead to the isolation of the molecular pathways that underlie the AD mechanisms. Thus, the set of mutations that are associated with a better or worse response to specific treatments may help the selection of the best drug for a specific patient, based on their personal genetic makeup. One of the most important theories of depression states that depressed mood is a consequence of a non-equilibrated monoaminergic tone. Consistent with this, evidence has shown that genes involved in the turnover of monoamines are modulators of the efficacy of AD pharmacotreatment. Serotonin, dopamine and norepinephrine are relevant monoamines, whose balance is thought to be disrupted during depressed disorder. Here, we will summarize current knowledge, listing the most relevant findings.

Monoamine Metabolic Enzymes
Tryptophan hydroxylase (TPH) catalyzes the rate-limiting step in serotonin biosynthesis, and exists in 2 isoforms: TPH1 and TPH2. TPH2 is brain specific [21, 22], and a correlation with depression and suicidal behavior has been reported [23, 24]. Tzvetkov et al. [25] reported that rs10897346 and rs1487278 located in this gene are associated with a favorable AD treatment outcome.

Catechol-O-methyltransferase (COMT) is involved in the catabolic pathways of norepinephrine and dopamine, a role associated with remarkable effects on the activity of ADs [26, 27]. A functional polymorphism consisting of a transition of guanine to adenine at codon 158 leads to a Val-Met substitution in membrane bound – COMT (and in position 108 in soluble – COMT) [28], which resulted in diminished activity of the protein [29, 30], a higher risk of suicidal behavior and personality traits [31], and a worse response to mirtazapine, paroxetine [32, 33], citalopram [26] and fluoxetine in male Asian patients [34]. Perlis et al. recently reported that rs165599 located in the COMT gene accounted for the 3% of variance in AD response after 6 weeks of treatment with duloxetine [35].

Monoamine oxidase A (MAO-A) catabolizes norepinephrine, dopamine and serotonin. A polymorphism located 1.2 kb upstream of the MAO-A coding sequences (variable number tandem repeat; VNTR) was reported to affect the transcription of the MAO-A promoter: alleles with 3.5 or 4 copies of the repeat sequence are transcribed 2–10 times more efficiently than those with 3 or 5 copies of the repeat [36]. Bipolar disorder – as well as suicidal behavior, personality features, aggressive behavior, alcoholism and AD response in females – has been associated with this and other polymorphisms in the MAO-A gene [37–43]. Moreover, it has been recently reported that the T941G polymorphism in the MAO-A gene is associated with mirtazapine response in females [44]. MAO and COMT determine the turnover of some relevant monoamines.

Monoamine Transporters
Other enzymes assist in the clearing of neurotransmitter from the synaptic cleft in a more specific way. The serotonin transporter clears serotonin from the neuronal surfaces, and it is the principal site of action of many ADs. Heils et al. [45] described an insertion deletion polymorphism in the promoter region of the gene able to impact the expression of the serotonin transporter: the long (l) allele has twice the expression in the basal state than the short (s) form. Moreover, numerous additional variants within the repetitive region occur [46] that are able to modify the expression of the gene [47, 48]. This variation was found to be associated with affective symptomatology (e.g. depression, bipolar disorder, anxiety disorders, eating disorders, substance abuse) and with pathological behaviors and personality traits related to anxiety, impulsivity and stress [49]. Moreover, the short allele was found to be associated with a worse response – including lack of effect and stronger side effects – to AD pharmacotreatment [50, 51]. This findings were found to be more consistent in Caucasian

samples, which could be due to a different prevalence of this mutation in diverse ethnicities. Another polymorphism influencing the expression of the serotonin transporter was identified by Ogilvie et al. [53] within intron 2 (STin2) and described as a 17-bp VNTR polymorphism. It has been associated with depressive disorder [52–54] and suicide behavior [55, 56]. STin2 also affected AD response [57, 58], although not consistently [51, 59, 60]. rs25531 was also found to impact the response to AD pharmacotreatment [61, 62].

Norepinephrine transporter clears norepinephrine from the synaptic cleft. A369P, F528C, G1287A, T128C and N292T were proved to be functional and impact the AD effect [63–65]. More recently, Uher et al. [66] reported that the variations rs60329 and rs1532701 located in this gene were associated with a favorable response to treatment with nortryptyline. However, the results are not unequivocal, and replication studies are warranted.

Finally, dopamine transporter ends the dopaminergic signal transmission. A 40-bp VNTR polymorphism in exon 15 of this gene that affects dopamine transporter expression [67] is associated with a faster onset of AD response when the allelic variant associated with enhanced expression (10 repeat variant) is present [68].

Monoamine Receptors

Not only the variations located in genes that regulate the turnover of monoamines were found to be associated with AD response: the sites of action of neurotransmitters may play a relevant role as well. Serotonin-1A receptors (5-HT_{1A}) play an inhibitory role in the serotonin system: its desensitization is thought to be one possible AD mechanism [69]. A common C(–1019)G single nucleotide polymorphism regulates its expression rate [70, 71] and impacts the efficacy of AD pharmacotreatment [60, 72–76]. Gly272Asp was found to modify the response to treatment as well [77]. Kato et al. [78] recently reported on the significant impact of rs6295. Serotonin-2A receptors (5-HT_{2A}) have an activating function in the central nervous system. There is a solid set of evidence that supports their involvement in the efficacy of AD treatments [79–81]. In particular, 5-HT_{2A} T(102)C and 5-HT_{2A} G(–1438)A were reported to be associated with the efficacy of the AD pharmacotreatment [82–86]. Perlis et al. [35] recently reported that variations rs9534505, rs1923884 and rs276035 were favorably associated with AD treatment outcome. Consistently, Uher et al. [66] reported that rs7324218, rs9316233 and rs2224721 were associated with response to AD pharmacotreatment. Serotonin-6 receptor (5-HT_6) is a G-protein coupled receptor which stimulates adenylyl cyclase via G(s) coupling together with 5-HT_4 and 5-HT_7. Recently, the involvement of this receptor in the AD mechanism has been reported [87, 88]; genetically, the variation T(267)C in the first exon may have a role in the modulation of AD response [89, 90]. A recently identified functional polymorphism in the β_1-adrenergic receptor G(1165)C leading to the amino acid variation Gly389Arg may play a functional role, resulting in a better and faster response to AD treatment [91]. Indeed, a recent report on a large sample of 873 major depressed patients failed to demonstrate the relevance of this gene in modulating the response to citalopram treatment [92]. Dopamine receptor D_2 is a G-protein-coupled receptor that inhibits adenylyl cyclase activity. A serine to cysteine change at codon 311 (DRD2 Ser311Cys) [93] may be functional, but showed no significant influence on AD response in some studies [94–96]. DRD4 has been extensively investigated as well, but no conclusive reports about its efficacy in modulating AD response can be formulated, due to the inconsistent findings in literature [94, 97, 98].

Other Relevant Genes

The monoaminergic theory of depression cannot be considered as comprehensive. Other theories sustain the role of chronic stress or immune deregulations as pivotal in determining depressive phenotypes. Consistently, corticotrophin-releasing hormone (CRH) receptor 1 is considered to play a key role in mediating the CRH-elicited effects in depression and anxiety [99]. CRH receptor 1 (CRHR1) antagonists have consistently demonstrated AD properties in experimental animals and humans [100–102]. Some evidence stands for a relevance of CRHR1 variants and AD response, in particular an association within rs242941 G/G genotype and homozygous GAG haplotype of the 3 single nucleotide polymorphisms and fluoxetine therapeutic response [103–104]. Citalopram treatment as well was found to be impacted by the presence of a variation (rs110402) located in this gene [105]. On the other hand, depressed individuals show impairment of their immune system [106–109], and interleukin-1 (IL-1) activity was showed to be altered in mood disorders [110–113]. Consistently, homozygosity for the –511T allele of the IL-1β gene was found to be associated with a trend towards less severe depressive symptoms and more favorable response to fluoxetine [114]. Finally, genetic variations located in genes that are involved in different systems have been reported to impact the effects of AD pharmacotreatment, including angiotensin-converting enzyme [115–118], the CLOCK gene [119–121] and the glutamatergic system [for a review, see 6].

In conclusion, there is a list of several candidate genes with a putative role in the regulation of AD response (table 2). Nevertheless, the lines of evidence that have been quoted here do not provide a conclusive picture of the pharmacodynamics of AD response, in that the replication rates are still too low with maybe the only exception of the short/long polymorphism of the serotonin transporter in Caucasian patients. The reasons for this substantial failure may rely on the complexity of the field which ranges from the phenotype definition to the intricate regulation of the genome [122, 123]. Nonetheless, even though pharmacogenetics appears to be still in its very infancy, it will likely help explaining how ADs exert their activity.

Table 2. Overview of genetic association studies on antidepressant efficacy and side effects

Candidate gene	Analyzed polymorphism	Number of studies[1]	Evidence
Serotonin transporter	44 bp Del/Ins (promoter)	45	+++++++++++++++++++++++++++++++++++++ – – – – – – – –
		8 (SE)	SE +++++ – – –
	VNTR Stin2	10	++++++ – – – –
		2 (SE)	SE + –
	rs25531	3	++ –
5-HT$_{1A}$ receptor	–1019C/G	11	++++++++ – – –
	272Gly/Asp	2	+ –
	rs10042486C/T	1	+
	rs1364043G/T	1	+
5-HT$_{2A}$ receptor	–1438G/A	7	+++ – – – –
		4 (SE)	SE +++ –
	102T/C	5	+ – – – –
		2 (SE)	SE + –
	–1420C/T	1	+
	rs7997012A/G	3	++ –
	rs6314C/T	2	+ –
	rs3125C/G	2	+ –
	rs1923882C/T	2	+ –
	rs1923884C/T	2	++
	rs9534505A/G	1	+
	rs2760351C/T	1	+
	rs9316233C/G	1	+
	rs2224721A/C	1	+
5-HT$_{3A}$ receptor	178C/T	3	+ SE – –
	195C/T	2	SE – –
5-HT$_{3B}$ receptor	129Tyr/Ser		SE + –
	–100 to –102AAG		+
	Del/Ins		SE + –
Tryptophan hydroxylase 1	218A/C	9	+++ – – – – – –
		1 (SE)	SE –
	–7180T/G	1	+
	–7065T/C	1	+
	–5806T/G	1	+
	rs1800532A/C	1	–
Tryptophan hydroxylase 2	rs1843809G/T	1	+
	rs1386492A/G	1	+
	rs1487276A/G	1	+
	rs10897346C/T	1	+
	rs1487278C/T	1	+
	rs2171363C/T	1	+
	rs1386494A/G	1	–
	1463G/A	1	– (treatment resistance)
	1487C/G	1	– (treatment resistance)
	1578T/G	1	– (treatment resistance)

Table 2 (continued)

Candidate gene	Analyzed polymorphism	Number of studies[1]	Evidence
Noradrenaline transporter (NET)	G1287A	2	++
	T-182C	1	+
	rs36029	1	–
	rs1532701	1	–
Dopamine transporter (DAT)	VNTR (40-bp in exon 15)	2	++
Monoamine oxidase A (MAO-A)	MAO-A-uVNTR	7	+++ – – – –
		1	SE +
	rs6323G/T	2	++
	rs1465108A/G	1	+
	rs6323A/C	1	+
	941T/G	1	+
Catechol-O-methyltransferase (COMT)	158Val/Met (472G/A)	7	++++++ –
	rs165599A/G	1	+
	rs165774A/G	1	+
	rs174696C/T	1	+
G-protein β-3 subunit	C825T	8	+++++ – – –
Brain-derived neurotrophic factor (BDNF)	196G/A (66Val/Met) (rs6265)	7[2]	++++ – – – –
	rs908867A/G	2	++
	rs61888800G/T	1	+
	rs7124442C/T	1[2]	+ –
	rs7103411C/T	1[2]	+ –
Glucocorticoid receptor-regulating co-chaperone of hsp-90 (FKBP5)	rs1360780 C/T	5[2]	+++ – – –
	rs3800373A/C	3[2]	++ – –
	rs4713916A/G	3[2]	++ – –
Angiotensin-converting enzyme (ACE)	ACE I/D	4	+++ –
Corticotrophin-releasing hormone receptor 1 (CRHR1)	rs1876828A/G	3	+ (rs242941)
	rs242939A/G	3	++ (haplotype of rs1876828, rs242939, rs24294)
	rs242941G/T	3	–
Dystrobrevin-binding-protein 1 (DTNBP1)	rs760761C/T	3	+ – –
	rs2619522A/C	2	+ –
	rs2005976A/G	1	+
	rs3213207G/A	1	–
	rs1011313A/G	1	–
	rs2619528A/G	1	–
Circadian locomotor output cycles kaput (CLOCK)	rs1801260 (T3111C)	2	+ –
	rs3736544A/G	1	+
	rs3749474C/T	1	+

Table 2 (continued)

Candidate gene	Analyzed polymorphism	Number of studies[1]	Evidence
Glycogen synthase kinase-3β (GSK-3β)	–50T/C (rs334558)	4	++++
	rs13321783C/T	1	+
	rs2319398G/T	1	+
	rs6808874A/T	1	–
Cyclic nucleotide phosphodiesterase (PDE)	PDE11A rs1880916	3	++ –
	PDE1A rs1549870	2	+ –
GRIK4	rs1954787	2	++
	rs12800734	1	+

We only summarized studies focusing on genes for which ≥2 independent studies observed significant associations, independent of further negative replications and position of the marker within the gene. Evidence of association (+) or no association (–) between polymorphism and treatment efficacy or side effects is indicated. 5-HT = Serotonin; uVNTR = upstream variable number of tandem repeats.

[1] Where not otherwise specified, this refers to those included for the association between polymorphism and treatment outcome; SE refers to number of studies included for association between gene variants and side effects.

[2] Study reporting double evidence for a single polymorphism due replication in >1 sample.

Conclusions and Perspectives

Pharmacogenetic studies of AD response have suggested several strong candidate genes involved in the pharmacokinetics and pharmacodynamics of these agents. However, comparisons across studies are complicated by a variety of critical methodological aspects such as differences in inclusion criteria, type of medication, outcome measure, evaluation of adverse effects, genetic coverage and ethnicity. Even when certain polymorphisms appear to show replicable association with AD treatment response, effect sizes across studies can be very different and it is difficult to discriminate whether the observed differences are chance findings or in fact related to clinical differences in the sample. Some genes also appear to specifically interfere with response to selected treatments, while others modulate response to various AD treatments, including non-pharmacological interventions. In summary, with more genome-wide association studies still outstanding, so far no clinically tested predictive markers have yet been established and larger more refined studies, at both phenotypic and genetic levels, are needed.

References

1 Murray CJ, Lopez AD: Evidence-based health policy: lessons from the Global Burden of Disease Study. Science 1996;274:740–743.
2 Stahl SM: Stahl's Essential Psychopharmacology, ed 3. Cambridge, Cambridge University Press, 2008.
3 Moncrieff J, Kirsch I: Efficacy of antidepressants in adults. BMJ 2005;331:551–557.
4 Nierenberg AA: Predictors of response to antidepressants general principals and clinical implications. Psychiatr Clin North Am 2003;26:345–352.
5 Perlis RH: Pharmacogenetic studies of antidepressant response: how far from the clinic? Psychiatr Clin North Am 2007;30:125–138.
6 Horstmann S, Binder EB: Pharmacogenomics of antidepressant drugs. Pharmacol Ther 2009;124:57–73.
7 Kirchheiner J, Nickchen K, Bauer M, Wong ML, Licinio J, Roots I, Brockmoller J: Pharmacogenetics of antidepressants and antipsychotics: the contribution of allelic variations to the phenotype of drug response. Mol Psychiatry 2004;9:442–473.
8 Ingelman-Sundberg M, Sim SC, Gomez A, Rodriguez-Antona C: Influence of cytochrome P450 polymorphism on drug therapies: pharmacogenetic, pharmacoepigenetic and clinical aspects. Pharmacol Ther 2007;116:496–526.
9 Spina E, Santoro V, D'Arrigo C: Clinically relevant pharmacokinetic drug interactions with second-generation antidepressants: an update. Clin Ther 2008;30:1206–1227.
10 Bertilsson L, Dahl ML, Dalen P, Al-Shurbaji A: Molecular genetics of CYP2D6: clinical relevance with focus on psychotropic drugs. Br J Clin Pharmacol 2002;53:111–122.
11 Bertilsson L: Metabolism of antidepressant and neuroleptic drugs by cytochrome P450s: clinical and interethnic aspects. Clin Pharmacol Ther 2007;82:606–609.
12 Wedlund PJ, de Leon J: Cytochrome P4502D6 and antidepressant toxicity and response: what is the evidence? Clin Pharmacol Ther 2004;75:373–375.
13 Kirchheiner J, Brosen K, Dahl ML, Gram LF, Kasper S, Roots I, Sjoqvist F, Spina E, Brockmoller J: CYP2D6 and CYP2C19 genotype-based dose recommendations for antidepressants: a first step towards subpopulation specific dosages. Acta Psychiatr Scand 2001;104:173–192.
14 de Leon J: The crucial role of therapeutic window in understanding the clinical relevance of the poor versus the ultrarapid metabolizer phenotype in subjects taking drugs metabolized by CYP2D6 and CYP2C19. J Clin Psychopharmacol 2007;27:241–245.
15 Callen DF, Baker E, Simmers RN, Seshadri R, Roninson IB: Localization of the human multiple drug resistance gene, MDR1, to 7q21.1. Hum Genet 1987;77:142–144.
16 Linnet K, Ejsing TB: A review on the impact of P-glycoprotein on the penetration of drugs into the brain: focus on psychotropic drugs. Eur Neuropsychopharmacol 2008;18:157–169.
17 Roberts RL, Joyce PR, Mulder RT, Begg EJ, Kennedy MA: A common P-glycoprotein polymorphism is associated with nortriptyline-induced postural hypotension in patients treated for major depression. Pharmacogenomics J 2002;2:191–196.
18 Laika B, Leucht S, Steimer W: ABCB1 (P-glycoprotein/MDR1) gene G2677T/a sequence variation (polymorphism): lack of association with side effects and therapeutic response in depressed inpatients treated with amitriptyline. Clin Chem 2006;52:893–895.
19 Fukui N, Suzuki Y, Sawamura K, Sugai T, Watanabe J, Inoue Y, Someya T: Dose-dependent effects of the 3435 C>T genotype of ABCB1 gene on the steady-state plasma concentration of fluvoxamine in psychiatric patients. Ther Drug Monit 2007;29:185–189.
20 Uhr M, Tontsch A, Namendorf C, Ripke S, Lucae S, Ising M, Dose T, Ebinger M, Rosenhagen M, Kohli M, Kloiber S, Salyakina D, Bettecken T, Specht M, Putz B, Binder EB, Muller-Myhsok B, Holsboer F: Polymorphisms in drug transporter gene ABCB1 predict antidepressant treatment response in depression. Neuron 2008;57:203–209.
21 Zill P, Buttner A, Eisenmenger W, Moller H, Ackenheil M, Bondy B: Analysis of tryptophan hydroxylase I and II mRNA expression in the human brain: a post-mortem study. J Psychiatr Res 2007;41:168–173.
22 Sakowski SA, Geddes TJ, Thomas DM, Levi E, Hatfield JS, Kuhn DM: Differential tissue distribution of tryptophan hydroxylase isoforms 1 and 2 as revealed with monospecific antibodies. Brain Res 2006;1085:11–18.
23 Zill P, Buttner A, Eisenmenger W, Moller HJ, Bondy B, Ackenheil M: Single nucleotide polymorphism and haplotype analysis of a novel tryptophan hydroxylase isoform (TPH2) gene in suicide victims. Biol Psychiatry 2004;56:581–586.
24 Zhou Z, Roy A, Lipsky R, Kuchipudi K, Zhu G, Taubman J, Enoch MA, Virkkunen M, Goldman D: Haplotype-based linkage of tryptophan hydroxylase 2 to suicide attempt, major depression, and cerebrospinal fluid 5-hydroxyindoleacetic acid in 4 populations. Arch Gen Psychiatry 2005;62:1109–1118.

25 Tzvetkov MV, Brockmoller J, Roots I, Kirchheiner J: Common genetic variations in human brain-specific tryptophan hydroxylase-2 and response to antidepressant treatment. Pharmacogenet Genomics 2008;18:495–506.

26 Arias B, Serretti A, Lorenzi C, Gasto C, Catalan R, Fananas L: Analysis of COMT gene (Val 158 Met polymorphism) in the clinical response to SSRIs in depressive patients of European origin. J Affect Disord 2006;90:251–256.

27 Mossner R, Simantov R, Marx A, Lesch KP, Seif I: Aberrant accumulation of serotonin in dopaminergic neurons. Neurosci Lett 2006;401:49–54.

28 Lachman HM, Morrow B, Shprintzen R, Veit S, Parsia SS, Faedda G, Goldberg R, Kucherlapati R, Papolos DF: Association of codon 108/158 catechol-O-methyltransferase gene polymorphism with the psychiatric manifestations of velo-cardio-facial syndrome. Am J Med Genet 1996;67:468–472.

29 Mannisto P, Kaakkola S: Catechol-O-methyltransferase (COMT): biochemistry, molecular biology, pharmacology, and clinical efficacy of the new selective COMT inhibitors. Pharmacol Rev 2005;51: 593–628.

30 Weinshilboum RM, Otterness DM, Szumlanski CL: Methylation pharmacogenetics: catechol O-methyltransferase, thiopurine methyltransferase, and histamine N-methyltransferase. Annu Rev Pharmacol Toxicol 1999;39:19–52.

31 Craddock N, O'Donovan MC, Owen MJ: Genes for schizophrenia and bipolar disorder? Implications for psychiatric nosology. Schizophr Bull 2006;32: 9–16.

32 Szegedi A, Rujescu D, Tadic A, Muller MJ, Kohnen R, Stassen HH, Dahmen N: The catechol-O-methyltransferase Val108/158Met polymorphism affects short-term treatment response to mirtazapine, but not to paroxetine in major depression. Pharmacogenomics J 2005;5:49–53.

33 Benedetti F, Colombo C, Pirovano A, Marino E, Smeraldi E: The catechol-O-methyltransferase Val(108/158)Met polymorphism affects antidepressant response to paroxetine in a naturalistic setting. Psychopharmacology (Berl) 2009;203:155–160.

34 Tsai SJ, Gau YT, Hong CJ, Liou YJ, Yu YW, Chen TJ: Sexually dimorphic effect of catechol-O-methyltransferase val158met polymorphism on clinical response to fluoxetine in major depressive patients. J Affect Disord 2009;113:183–187.

35 Perlis RH, Fijal B, Adams DH, Sutton VK, Trivedi MH, Houston JP: Variation in catechol-O-methyltransferase is associated with duloxetine response in a clinical trial for major depressive disorder. Biol Psychiatry 2009;65:785–791.

36 Sabol SZ, Hu S, Hamer D: A functional polymorphism in the monoamine oxidase A gene promoter. Human Genetic 1998;103:273–279.

37 Newman TK, Syagailo YV, Barr CS, Wendland JR, Champoux M, Graessle M, Suomi SJ, Higley JD, Lesch KP: Monoamine oxidase A gene promoter variation and rearing experience influences aggressive behavior in rhesus monkeys. Biol Psychiatry 2005;57:167–172.

38 Manuck SB, Flory JD, Ferrell RE, Mann JJ, Muldoon MF: A regulatory polymorphism of the monoamine oxidase-A gene may be associated with variability in aggression, impulsivity, and central nervous system serotonergic responsivity. Psychiatry Res 2000;95:9–23.

39 Preisig M, Ferrero F, Malafosse A: Monoamine oxidase A and tryptophan hydroxylase gene polymorphisms: are they associated with bipolar disorder? Am J Pharmacogenomics 2005;5:45–52.

40 Bondy B, Buettner A, Zill P: Genetics of suicide. Mol Psychiatry 2006;11:336–351.

41 Serretti A, Mandelli L, Lorenzi C, Landoni S, Calati R, Insacco C, Cloninger CR: Temperament and character in mood disorders: influence of DRD4, SERTPR, TPH and MAO-A polymorphisms. Neuropsychobiology 2006;53:9–16.

42 Yu Y, Tsai S, Hong C, Chen T, Chen M, Yang C: Association study of a monoamine oxidase a gene promoter polymorphism with major depressive disorder and antidepressant response. Neuropsychopharmacology 2005;30:1719–1723.

43 Parsian A: Sequence analysis of exon 8 of MAO-A gene in alcoholics with antisocial personality and normal controls. Genomics 1999;55:290–295.

44 Tadic A, Muller MJ, Rujescu D, Kohnen R, Stassen HH, Dahmen N, Szegedi A: The MAOA T941G polymorphism and short-term treatment response to mirtazapine and paroxetine in major depression. Am J Med Genet B Neuropsychiatr Genet 2007;144: 325–331.

45 Heils A, Teufel A, Petri S, Stöber G, Riederer P, Bengel D, Lesch P: Allelic variation of human serotonin transporter gene expression. J Neurochem 1996;66: 2621–2624.

46 Nakamura M, Ueno S, Sano A, Tanabe H: The human serotonin transporter gene linked polymorphism (5-HTTLPR) shows ten novel allelic variants. Mol Psychiatry 2000;5:32–38.

47 Hu X, Oroszi G, Chun J, Smith TL, Goldman D, Schuckit MA: An expanded evaluation of the relationship of four alleles to the level of response to alcohol and the alcoholism risk. Alcohol Clin Exp Res 2005;29:8–16.

48 Hu XZ, Rush AJ, Charney D, Wilson AF, Sorant AJ, Papanicolaou GJ, Fava M, Trivedi MH, Wisniewski SR, Laje G, Paddock S, McMahon FJ, Manji H, Lipsky RH: Association between a functional serotonin transporter promoter polymorphism and citalopram treatment in adult outpatients with major depression. Arch Gen Psychiatry 2007;64:783–792.

49 Serretti A, Calati R, Mandelli L, De Ronchi D: Serotonin transporter gene variants and behavior: a comprehensive review. Curr Drug Targets 2006;7: 1659–1669.

50 Serretti A, Mandelli L, Lorenzi C, Pirovano A, Olgiati P, Colombo C, Smeraldi E: Serotonin transporter gene influences the time course of improvement of 'core' depressive and somatic anxiety symptoms during treatment with SSRIs for recurrent mood disorders. Psychiatry Res 2007;149:185–193.

51 Smits K, Smits L, Peeters F, Schouten J, Janssen R, Smeets H, van Os J, Prins M: Serotonin transporter polymorphisms and the occurrence of adverse events during treatment with selective serotonin reuptake inhibitors. Int Clin Psychopharmacol 2007;22:137–143.

52 MacKenzie A, Quinn J: A serotonin transporter gene intron 2 polymorphic region, correlated with affective disorders, has allele-dependent differential enhancer-like properties in the mouse embryo. Proc Natl Acad Sci USA 1999;96:15251–15255.

53 Ogilvie AD, Battersby S, Bubb VJ, Fink G, Harmar AJ, Goodwin GM, Smith CAD: Polymorphism in serotonin transporter gene associated with susceptibility to major depression. Lancet 1996;347:731–733.

54 Gutierrez B, Pintor L, Gasto C, Rosa A, Bertranpetit J, Vieta E, Fananas L: Variability in the serotonin transporter gene and increased risk for major depression with melancholia. Hum Gen 1998;103: 319–322.

55 Gaysina D, Zainullina A, Gabdulhakov R, Khusnutdinova E: The serotonin transporter gene: polymorphism and haplotype analysis in Russian suicide attempters. Neuropsychobiology 2006;54:70–74.

56 Lopez de Lara C, Dumais A, Rouleau G, Lesage A, Dumont M, Chawky N, Alda M, Benkelfat C, Turecki G: STin2 variant and family history of suicide as significant predictors of suicide completion in major depression. Biol Psychiatry 2006;59:114–120.

57 Kim DK, Lim SW, Lee S, Sohn SE, Kim S, Hahn CG, Carroll BJ: Serotonin transporter gene polymorphism and antidepressant response. Neuroreport 2000;11:215–219.

58 Smits KM, Smits LJ, Schouten JS, Stelma FF, Nelemans P, Prins MH: Influence of SERTPR and STin2 in the serotonin transporter gene on the effect of selective serotonin reuptake inhibitors in depression: a systematic review. Mol Psychiatry 2004;9: 433–441.

59 Ito K, Yoshida K, Sato K, Takahashi H, Kamata M, Higuchi H, Shimizu T, Itoh K, Inoue K, Tezuka T, Suzuki T, Ohkubo T, Sugawara K, Otani K: A variable number of tandem repeats in the serotonin transporter gene does not affect the antidepressant response to fluvoxamine. Psychiatry Res 2002;111: 235–239.

60 Hong CJ, Chen TJ, Yu YW, Tsai SJ: Response to fluoxetine and serotonin 1A receptor (C-1019G) polymorphism in Taiwan Chinese major depressive disorder. Pharmacogenomics J 2006;6:27–33.

61 Kraft J, Slager S, McGrath P, Hamilton S: Sequence analysis of the serotonin transporter and associations with antidepressant response. Biol Psychiatry 2005;58:374–381.

62 Smeraldi E, Serretti A, Artioli P, Lorenzi C, Catalano M: Serotonin transporter gene-linked polymorphic region: possible pharmacogenetic implications of rare variants. Psychiatr Genet 2006;16: 153–158.

63 Hahn MK, Mazei-Robison MS, Blakely RD: Single nucleotide polymorphisms in the human norepinephrine transporter gene affect expression, trafficking, antidepressant interaction, and protein kinase C regulation. Mol Pharmacol 2005;68:457–466.

64 Kim H, Lim SW, Kim S, Kim JW, Chang YH, Carroll BJ, Kim DK: Monoamine transporter gene polymorphisms and antidepressant response in Koreans with late-life depression. JAMA 2006;296: 1609–1618.

65 Yoshida K, Takahashi H, Higuchi H, Kamata M, Ito K, Sato K, Naito S, Shimizu T, Itoh K, Inoue K, Suzuki T, Nemeroff CB: Prediction of antidepressant response to milnacipran by norepinephrine transporter gene polymorphisms. Am J Psychiatry 2004; 161:1575–1580.

66 Uher R, Huezo-Diaz P, Perroud N, Smith R, Rietschel M, Mors O, Hauser J, Maier W, Kozel D, Henigsberg N, Barreto M, Placentino A, Dernovsek MZ, Schulze TG, Kalember P, Zobel A, Czerski PM, Larsen ER, Souery D, Giovannini C, Gray JM, Lewis CM, Farmer A, Aitchison KJ, McGuffin P, Craig I: Genetic predictors of response to antidepressants in the GENDEP project. Pharmacogenomics J 2009;9:225–233.

67 Fuke S, Suo S, Takahashi N, Koike H, Sasagawa N, Ishiura S: The VNTR polymorphism of the human dopamine transporter (DAT1) gene affects gene expression. Pharmacogenomics J 2001;1:152–156.

68 Kirchheiner J, Nickchen K, Sasse J, Bauer M, Roots I, Brockmoller J: A 40-basepair VNTR polymorphism in the dopamine transporter (DAT1) gene and the rapid response to antidepressant treatment. Pharmacogenomics J 2007;7:48–55.

69 Perez V, Gilaberte I, Faries D, Alvarez E, Artigas F: Randomised, double-blind, placebo-controlled trial of pindolol in combination with fluoxetine antidepressant treatment. Lancet 1997;349:1594–1597.

70 Lemonde S, Turecki G, Bakish D, Du L, Hrdina P, Bown C, Sequeira A, Kushwaha N, Morris S, Basak A, Ou X, Albert P: Impaired trans-repression at a 5-HT1A receptor gene polymorphism associated with major depression and suicide. J Neurosci 2003; 23:8788–8799.

71 Albert PR, Lemonde S: 5-HT1A receptors, gene repression, and depression: guilt by association. Neuroscientist 2004;10:575–593.
72 Lemonde S, Du L, Bakish D, Hrdina P, Albert PR: Association of the C(1019)G 5-HT1A functional promoter polymorphism with antidepressant response. Int J Neuropsychopharmacol 2004;7:501–506.
73 Serretti A, Artioli P, Lorenzi C, Pirovano A, Tubazio V, Zanardi R: The C(-1019)G polymorphism of the 5-HT1A gene promoter and antidepressant response in mood disorders: preliminary findings. Int J Neuropsychopharmacol 2004;7:453–460.
74 Arias B, Catalan R, Gasto C, Gutierrez B, Fananas L: Evidence for a combined genetic effect of the 5-HT1A receptor and serotonin transporter genes in the clinical outcome of major depressive patients treated with citalopram. J Psychopharmacol 2005; 19:166–172.
75 Baune BT, Hohoff C, Roehrs T, Deckert J, Arolt V, Domschke K: Serotonin receptor 1A-1019C/G variant: impact on antidepressant pharmacoresponse in melancholic depression? Neurosci Lett 2008;436: 111–115.
76 Yu YW, Tsai SJ, Liou YJ, Hong CJ, Chen TJ: Association study of two serotonin 1A receptor gene polymorphisms and fluoxetine treatment response in Chinese major depressive disorders. Eur Neuropsychopharmacol 2006;16:498–503.
77 Suzuki Y, Sawamura K, Someya T: The effects of a 5-hydroxytryptamine 1A receptor gene polymorphism on the clinical response to fluvoxamine in depressed patients. Pharmacoeconomics J 2004;4: 283–286.
78 Kato M, Fukuda T, Wakeno M, Okugawa G, Takekita Y, Watanabe S, Yamashita M, Hosoi Y, Azuma J, Kinoshita T, Serretti A: Effect of 5-HT1A gene polymorphisms on antidepressant response in major depressive disorder. Am J Med Genet B Neuropsychiatr Genet 2009;150:115–123.
79 Maj J, Bijak M, Dziedzicka-Wasylewska M, Rogoz R, Rogoz Z, Skuza G, Tokarski T: The effects of paroxetine given repeatedly on the 5-HT receptor subpopulations in the rat brain. Psychopharmacology (Berl) 1996;127:73–82.
80 Meyer JH, Kapur S, Eisfeld B, Brown GM, Houle S, DaSilva J, Wilson AA, Rafi-Tari S, Mayberg HS, Kennedy SH: The effect of paroxetine on 5-HT2A receptors in depression: an [^{18}F]setoperone PET imaging study. Am J Psychiatry 2001;158:78–85.
81 Hemrick-Luecke SK, Snoddy HD, Fuller RW: Evaluation of nefazodone as a serotonin uptake inhibitor and a serotonin antagonist in vivo. Life Sci 1994; 55:479–483.
82 Murphy GM Jr, Kremer C, Rodrigues HE, Schatzberg AF: Pharmacogenetics of antidepressant medication intolerance. Am J Psychiatry 2003;160: 1830–1835.
83 Choi M, Kang R, Ham B, Jeong H, Lee M: Serotonin receptor 2A gene polymorphism (–1438A/G) and short-term treatment response to citalopram. Neuropsychobiology 2005;52:155–162.
84 Suzuki Y, Sawamura K, Someya T: Polymorphisms in the 5-hydroxytryptamine 2A receptor and CytochromeP4502D6 genes synergistically predict fluvoxamine-induced side effects in Japanese depressed patients. Neuropsychopharmacology 2006; 31:825–831.
85 Bishop JR, Moline J, Ellingrod VL, Schultz SK, Clayton AH: Serotonin 2A –1438 G/A and G-protein Beta3 subunit C825T polymorphisms in patients with depression and SSRI-associated sexual side-effects. Neuropsychopharmacology 2006;31: 2281–2288.
86 Kato M, Fukuda T, Wakeno M, Fukuda K, Okugawa G, Ikenaga Y, Yamashita M, Takekita Y, Nobuhara K, Azuma J, Kinoshita T: Effects of the serotonin type 2A, 3A and 3B receptor and the serotonin transporter genes on paroxetine and fluvoxamine efficacy and adverse drug reactions in depressed Japanese patients. Neuropsychobiology 2006;53:186–195.
87 Svenningsson P, Tzavara ET, Qi H, Carruthers R, Witkin JM, Nomikos GG, Greengard P: Biochemical and behavioral evidence for antidepressant-like effects of 5-HT6 receptor stimulation. J Neurosci 2007;27:4201–4209.
88 Wesolowska A, Nikiforuk A: Effects of the brain-penetrant and selective 5-HT6 receptor antagonist SB-399885 in animal models of anxiety and depression. Neuropharmacology 2007;52:1274–1283.
89 Kohen R, Metcalf MA, Khan N, Druck T, Huebner K, Lachowicz JE, Meltzer HY, Sibley DR, Roth BL, Hamblin MW: Cloning, characterization, and chromosomal localization of a human 5-HT6 serotonin receptor. J Neurochem 1996;66:47–56.
90 Lee S, Lee K, Lee H, Ham B, Ryu S, Lee M: Association between the 5-HT6 receptor C267T polymorphism and response to antidepressant treatment in major depressive disorder. Psychiatry Clin Neurosci 2005;59:140–145.
91 Zill P, Baghai TC, Engel R, Zwanzger P, Schule C, Minov C, Behrens S, Bottlender R, Jager M, Rupprecht R, Moller HJ, Ackenheil M, Bondy B: Beta-1-adrenergic receptor gene in major depression: influence on antidepressant treatment response. Am J Med Genet 2003;120B:85–89.
92 Crowley JJ, Lipsky RH, Lucki I, Berrettini WH: Variation in the genes encoding vesicular monoamine transporter 2 and beta-1 adrenergic receptor and antidepressant treatment outcome. Psychiatr Genet 2008;18:248–251.
93 Itokawa M, Arinami T, Futamura N, Hamaguchi H, Toru M: A structural polymorphism of human dopamine D_2 receptor, D_2(Ser311→Cys). Biochem Biophys Res Commun 1993;196:1369–1375.

94 Serretti A, Zanardi R, Cusin C, Rossini D, Lilli R, Lorenzi C, Lattuada E, Smeraldi E: No association between dopamine D_2 and D_4 receptor gene variants and antidepressant activity of two selective serotonin reuptake inhibitors. Psychiatry Res 2001;104:195–203.

95 Benedetti F, Serretti A, Colombo C, Lilli R, Lorenzi C, Smeraldi E: Dopamine receptor D_2 and D_3 gene variants are not associated with the antidepressant effect of total sleep deprivation in bipolar depression. Psychiatry Res 2003;118:241–247.

96 Serretti A, Artioli P, Zanardi R, Lorenzi C, Rossini D, Cusin C, Arnoldi A, Catalano M: Genetic features of antidepressant induced mania and hypomania in bipolar disorder. Psychopharmacology (Berl) 2004;174:504–511.

97 Serretti A, Benedetti F, Colombo C, Lilli R, Lorenzi C, Smeraldi E: Dopamine receptor D_4 is not associated with antidepressant activity of sleep deprivation. Psychiatry Res 1999;89:107–114.

98 Garriock HA, Delgado P, Kling MA, Carpenter LL, Burke M, Burke WJ, Schwartz T, Marangell LB, Husain M, Erickson RP, Moreno FA: Number of risk genotypes is a risk factor for major depressive disorder: a case control study. Behav Brain Funct 2006;2:24.

99 Van Pett K, Viau V, Bittencourt JC, Chan RK, Li HY, Arias C, Prins GS, Perrin M, Vale W, Sawchenko PE: Distribution of mRNAs encoding CRF receptors in brain and pituitary of rat and mouse. J Comp Neurol 2000;428:191–212.

100 Seymour P, Schmidt A, Schulz D: The pharmacology of CP-154,526, a non-peptide antagonist of the CRH1 receptor: a review. CNS Drug Rev 2003; 9:57–96.

101 Overstreet D, Griebel G: Antidepressant-like effects of CRF1 receptor antagonist SSR125543 in an animal model of depression. Eur J Pharmacol 2004;497:49–53.

102 Kehne JH: The CRF1 receptor, a novel target for the treatment of depression, anxiety, and stress-related disorders. CNS Neurol Disord Drug Targets 2007;6:163–182.

103 Licinio J, O'Kirwan F, Irizarry K, Merriman B, Thakur S, Jepson R, Lake S, Tantisira K, Weiss S, Wong M: Association of a corticotropin-releasing hormone receptor 1 haplotype and antidepressant treatment response in Mexican-Americans. Mol Psychiatry 2004;9:1075–1082.

104 Liu Z, Zhu F, Wang G, Xiao Z, Tang J, Liu W, Wang H, Liu H, Wang X, Wu Y, Cao Z, Li W: Association study of corticotropin-releasing hormone receptor1 gene polymorphisms and antidepressant response in major depressive disorders. Neurosci Lett 2007;414:155–158.

105 Papiol S, Arias B, Gasto C, Gutierrez B, Catalan R, Fananas L: Genetic variability at HPA axis in major depression and clinical response to antidepressant treatment. J Affect Disord 2007;104:83–90.

106 Dentino AN, Pieper CF, Rao MK, Currie MS, Harris T, Blazer DG, Cohen HJ: Association of interleukin-6 and other biologic variables with depression in older people living in the community. J Am Geriatr Soc 1999;47:6–11.

107 Herbert TB, Cohen S: Depression and immunity: a meta-analytic review. Psychol Bull 1993;113: 472–486.

108 Kiecolt-Glaser JK, Glaser R: Depression and immune function: central pathways to morbidity and mortality. J Psychosom Res 2002;53:873–876.

109 Zorrilla EP, Luborsky L, McKay JR, Rosenthal R, Houldin A, Tax A, McCorkle R, Seligman DA, Schmidt K: The relationship of depression and stressors to immunological assays: a meta-analytic review. Brain Behav Immun 2001;15:199–226.

110 Maes M, Bosmans E, Meltzer HY, Scharpe S, Suy E: Interleukin-1 beta: a putative mediator of HPA axis hyperactivity in major depression? Am J Psychiatry 1993;150:1189–1193.

111 Anisman H, Ravindran AV, Griffiths J, Merali Z: Interleukin-1 beta production in dysthymia before and after pharmacotherapy. Biol Psychiatry 1999;46:1649–1655.

112 Anisman H, Kokkinidis L, Merali Z: Further evidence for the depressive effects of cytokines: anhedonia and neurochemical changes. Brain Behav Immun 2002;16:544–556.

113 Licinio J, Wong ML: The role of inflammatory mediators in the biology of major depression: central nervous system cytokines modulate the biological substrate of depressive symptoms, regulate stress-responsive systems, and contribute to neurotoxicity and neuroprotection. Mol Psychiatry 1999;4:317–327.

114 Yu YW, Chen TJ, Hong CJ, Chen HM, Tsai SJ: Association study of the interleukin-1beta (C-511T) genetic polymorphism with major depressive disorder, associated symptomatology, and antidepressant response. Neuropsychopharmacology 2003;28:1182–1185.

115 Baghai TC, Schule C, Zwanzger P, Minov C, Schwarz MJ, de Jonge S, Rupprecht R, Bondy B: Possible influence of the insertion/deletion polymorphism in the angiotensin I-converting enzyme gene on therapeutic outcome in affective disorders. Mol Psychiatry 2001;6:258–259.

116 Baghai T, Schule C, Zwanzger P, Zill P, Ella R, Eser D, Deiml T, Minov C, Rupprecht R, Bondy B: Influence of a functional polymorphism within the angiotensin I-converting enzyme gene on partial sleep deprivation in patients with major depression. Neurosci Lett 2003;339:223–226.

117 Baghai TC, Schule C, Zill P, Deiml T, Eser D, Zwanzger P, Ella R, Rupprecht R, Bondy B: The angiotensin I converting enzyme insertion/deletion polymorphism influences therapeutic outcome in major depressed women, but not in men. Neurosci Lett 2004;363:38–42.
118 Bondy B, Baghai T, Zill P, Schule C, Eser D, Deiml T, Zwanzger P, Ella R, Rupprecht R: Genetic variants in the angiotensin I-converting-enzyme (ACE) and angiotensin II receptor (AT1) gene and clinical outcome in depression. Prog Neuropsychopharmacol Biol Psychiatry 2005;29:1094–1099.
119 Benedetti F, Serretti A, Colombo C, Barbini B, Lorenzi C, Campori E, Smeraldi E: Influence of CLOCK gene polymorphism on circadian mood fluctuation and illness recurrence in bipolar depression. Am J Med Genet 2003;123B:23–26.
120 Serretti A, Benedetti F, Mandelli L, Lorenzi C, Pirovano A, Colombo C, Smeraldi E: Genetic dissection of psychopathological symptoms: insomnia in mood disorders and CLOCK gene polymorphism. Am J Med Genet 2003;121B:39–43.
121 Serretti A, Cusin C, Benedetti F, Mandelli L, Pirovano A, Zanardi R, Colombo C, Smeraldi E: Insomnia improvement during antidepressant treatment and CLOCK gene polymorphism. Am J Med Genet B Neuropsychiatr Genet 2005; 137B:36–39.
122 Serretti A, Calati R, Oasi O, De Ronchi D, Colombo C: Dissecting the determinants of depressive disorders outcome: an in depth analysis of two clinical cases. Ann Gen Psychiatry 2007;6:5.
123 Drago A, De Ronchi D, Serretti A: Incomplete coverage of candidate genes: a poorly considered bias. Curr Genomics 2007;8:476–483.

Prof. Edoardo Spina
Department of Clinical and Experimental Medicine and Pharmacology
Section of Pharmacology, University of Messina, Policlinico Universitario
Via Consolare Valeria
IT–98125 Messina (Italy)
Tel. +39 090 221 3647, Fax +39 090 221 3300, E-Mail espina@unime.it

Schwab M, Kaschka WP, Spina E (eds): Pharmacogenomics in Psychiatry.
Adv Biol Psychiatry. Basel, Karger, 2010, vol 25, pp 75–102

Pharmacogenomics of Attention-Deficit/Hyperactivity Disorder

Tanya Froehlich[a] · Mark A. Stein[b]

[a]Division of Developmental and Behavioral Pediatrics, Department of Pediatrics, Cincinnati Children's Hospital Medical Center, Cincinnati, Ohio, and [b]Department of Psychiatry and HALP Clinic/ADHD Research Center, University of Illinois at Chicago, Chicago, Ill., USA

Abstract

Attention-deficit/hyperactivity disorder (ADHD) is a promising disorder for pharmacogenetic and pharmacogenomic studies due to the high heritability of ADHD, as well as the significant variability in ADHD medication dosing, duration of effect, efficacy, and tolerability. This article summarizes ADHD pharmacogenetic investigations to date, which have primarily focused on response to methylphenidate. The most well-studied genes in ADHD pharmacogenetic studies are the dopamine transporter and dopamine receptor D4. Additional genes that have been associated with methylphenidate response include the adrenergic α2A receptor, carboxylesterase 1, catechol-O-methyltransferase, dopamine receptor D5, norepinephrine transporter protein 1, serotonin transporter, and synaptosomal-associated protein 25. Unfortunately, the results of current ADHD pharmacogenetic studies have not been entirely consistent, possibly due to small sample sizes and differences in study design, medication dosing regimens, and outcome measures. At present, researchers are increasingly interested in going beyond individual candidate genes to investigate gene-gene and gene-environment interactions or pathways, as well as whole-genome approaches. Potential clinical applications may include the development of treatment efficacy and side effect prediction algorithms that incorporate the interplay of genetic and environmental factors, as well as the future development of more targeted treatment strategies.

Attention-deficit/hyperactivity disorder (ADHD) is a prevalent psychiatric disorder characterized by difficulties with attention, impulsivity, and/or overactivity [1], and associated with impaired social, academic, adaptive, and occupational functioning [2, 3]. Stimulant medications rapidly improve symptoms in 50–75% of children with ADHD, although approximately 25% do not respond to or tolerate pharmacotherapy with one agent [4–6]. Even among responders, there is marked variability in optimal dosage, duration of effect, and tolerability [7], as well as poor adherence to treatment [8]. Unfortunately, few consistent predictors of ADHD medication efficacy, dose, or adverse effects have been identified [7]. As a result, treatment is determined empirically through a gradual dosage titration and a trial and error approach to different medication preparations [9].

The high heritability of ADHD is well-established [10]. The search for ADHD candidate genes was initially driven by the understanding that medications for the disorder target the catecholamine system [11, 12]. Presumably, variability in drug response may also be linked to genetic factors associated with relevant neural circuits [13]. As a result, there is considerable interest in the possibility of personalizing ADHD treatment based upon genetic characteristics that predict treatment response. Personalized ADHD treatment regimens have the potential to increase overall medication effectiveness by maximizing symptom reduction, lessening adverse effects, and improving long-term tolerability. However, ADHD pharmacogenetics/genomics is still a relatively young field, with further development necessary before research findings can be translated into clinical practice. This article summarizes current research findings, reviews previous studies' methodological limitations, and discusses potential clinical applications of ADHD pharmacogenetic and pharmacogenomic studies.

Definition of Terms: Pharmacogenetics and Pharmacogenomics

Pharmacogenetics is the study of genetic variability in medication response [14], and often focuses on large clinical effects of single gene variants. It is hoped that by studying individual candidate genes, susceptibility to adverse effects or non-response can be linked to specific gene variants that affect drug metabolizing enzymes, receptors, or transporters [15]. Pharmacogenomic approaches emphasize many genomic loci, including large biological pathways and the whole genome, to better understand and develop pharmacological treatments [16]. The hallmark of pharmacogenomics is the ability to study simultaneously the contribution to drug effects of many genes using genomic techniques [15]. Nonetheless, in practice the terms pharmacogenetics and pharmacogenomics are often used interchangeably [17].

Genetic Studies of ADHD Susceptibility

Candidate genes in ADHD susceptibility studies have largely been selected based on our understanding of stimulant medications' mechanisms of action [12], including blockade of the dopamine and norepinephrine transporters (SLC6A3 and SLC6A2, respectively), inhibition of monoamine oxidase (which metabolizes dopamine and norepinephrine), and enhanced release of the catecholamines from the presynaptic cell [18]. As a result, ADHD susceptibility studies have focused largely on catecholamine-related genes, although there is increasing interest in brain circuits related to additional neurotransmitters [19].

Candidate genes associated with increased risk of ADHD based on pooled odds ratios across 3 or more studies are the dopamine receptors (DRD4 and DRD5), dopamine transporter (SLC6A3), dopa-β-hydroxylase (DBH), serotonin receptor (HTR1B), serotonin transporter (SLC6A4), and synaptosomal-associated protein 25 (SNAP25) [12]. Other genes of increasing interest in ADHD susceptibility studies include catechol-O-methyltransferase (COMT) [20, 21], the adrenergic α2A receptor (ADRA2A) [22–24], and the norepinephrine transporter (SLC6A2) [25, 26].

ADHD Pharmacogenetic Research Studies

While knowledge about ADHD medications' mechanisms of action initially informed the search for genes related to increased risk for the disorder, these same polymorphisms are logical candidates to predict medication response [27]. To date, the majority of pharmacogenetic studies have examined methylphenidate (MPH) response. Several catecholamine-related candidate genes have been associated with treatment effects on ADHD symptoms (tables 1, 2). However, as yet, few consistent findings have emerged, and the nature, magnitude, and direction of purported genetic effects remain unclear. Small sample sizes and variations in study design (open studies vs. randomized controlled trials) may be partially responsible for the disparate findings [28]. Concerns raised regarding retrospective and naturalistic observational studies include potential biased ascertainment of outcome [29, 30], as well as a tendency to underestimate environmental contributions and overestimate genetic effects [31]. Therefore, we present results separately for placebo-controlled trials and naturalistic studies, first summarizing studies of the dopamine transporter (SLC6A3) and DRD4, the most well-studied genes in ADHD pharmacogenetic studies (fig. 1). We then review findings for additional genes that have been significantly associated with stimulant medication effects on ADHD symptoms in at least one prior published study, including the ADRA2A receptor, carboxylesterase 1 (CES1), COMT, DRD5, norepinephrine transporter protein 1 (SLC6A2), serotonin transporter (SLC6A4), and SNAP25. Other genes that have been evaluated in at least one ADHD pharmacogenetic study but have not shown evidence of significant main effects are referenced in table 1, including the dopamine receptor D2 (DRD2), nicotinic acetylcholine α4-receptor (CHRNA4), and serotonin receptors 1B and 2A (HTR1B, HTR2A). Studies cited in this review were identified using the PubMed [32] and PsycINFO [33] databases, in addition to references from relevant papers ascertained during the database search.

Dopamine Transporter (SLC6A3)

SLC6A3 encodes a presynaptic protein responsible for dopamine reuptake from the synapse. Stimulant medications inhibit SLC6A3, thereby increasing synaptic dopa-

Table 1. Pharmacogenetic studies of methylphenidate effects on ADHD symptoms in children and adolescents

Gene	First author	Design	Sample size	Study location	Outcome
Adrenergic α2A receptor (ADRA2A)	Polanczyk [74]	prospective, open-label	106	Brazil	improved effects on inattention symptoms with G allele
	da Silva [75]	prospective, open-label	59	Brazil	improved effects on inattention symptoms with G allele
	Cheon [76]	prospective, open-label	114	South Korea	greater rates of 'good response' to MPH and improved effects on total ADHD symptom score with homozygous G allele
Carboxyl-esterase 1 (CES1)	Nemoda [79]	prospective, open-label	122	Hungary	Gly homozygotes required higher MPH doses for symptom reduction
Catechol-O-methyl-transferase (COMT)	McGough [42]	prospective, double-blind, placebo-controlled, dose response	82	USA (California)	trend toward diminished effects on ADHD total symptoms score with homozygous val allele
	Kereszturi [49]	prospective, open-label	122	Hungary	improved effects on hyperactive-impulsive symptoms with homozygous val allele
	Cheon [82]	prospective, open-label	124	South Korea	trend toward greater MPH effects on symptoms with homozygous val allele; significant association between MPH non-response and homozygous met allele
Dopamine transporter (SLC6A3)	Stein [39]	prospective, double-blind, placebo-controlled, dose response	47	USA (Washington, DC)	different dose-response curves by SLC6A3 genotype, decreased effects on ADHD symptoms with homozygous 9-repeat
	McGough [41]	prospective, double-blind, placebo-controlled	81 (pre-schoolers)	USA (6 sites)	decreased effects with homozygous 9-repeat on parent ratings, but no effect on parent-teacher composite ratings
	Joober [40]	prospective, double-blind, placebo-controlled	159	Canada	decreased effects with homozygous 9-repeat on parent ratings
	McGough [42]	prospective, double-blind, placebo-controlled, dose response	82	USA (California)	no effect of 10-repeat allele on ADHD symptoms, effects of homozygous 9-repeat not assessed
	Winsberg [38]	prospective, open-label	30	USA (New York)	decreased effects with homozygous 10-repeat
	Roman [47]	prospective, open-label	50	Brazil	decreased effects with homozygous 10-repeat
	Kirley [45]	retrospective report	119	Ireland	increased effects with number of 10-repeats
	Cheon [46]	prospective, open-label with SPECT imaging	11	Korea	decreased effects with homozygous 10-repeat
	Langley [54]	retrospective report	168	UK (Wales)	no effect on ADHD symptoms
	Van der Meulen [52]	retrospective report	82	The Nether-lands	no effect on ADHD symptoms
	Bellgrove [50]	prospective, open label	26	Ireland	no effect of the 10-repeat allele or the 10/3 haplotype on ADHD symptoms

Table 1 (continued)

Gene	First author	Design	Sample size	Study location	Outcome
	Zeni [53]	prospective, open label	111	Brazil	no effect on ADHD symptoms
	Keresztúri [49]	prospective, open label	122	Hungary	no effect on ADHD symptoms
	Purper-Ouakil [48]	prospective, open label	141	France	decreased effects with homozygous 10-repeat
	Tharoor [51]	retrospective report	156	USA (Missouri)	no effect on ADHD symptoms
Dopamine receptor (DRD2)	Winsburg [38]	prospective, open-label	30	USA (New York)	no effect on ADHD symptoms
Dopamine receptor (DRD4)	McGough [41]	prospective, double-blind, placebo controlled	81 (pre-schoolers)	USA (6 sites)	no effect of 7-repeat allele on ADHD symptoms; promoter polymorphism (240-bp homozygotes) associated with improved effects on ADHD symptoms
	McGough [42]	prospective, double-blind, placebo-controlled, dose response	82	USA (California)	no effect on ADHD symptoms. 4-repeat VNTR and promoter polymorphisms associated with significant gene x MPH dose joint effects on math performance
	Winsburg [38]	prospective, open-label	30	USA (New York)	no effect of 7-repeat allele on ADHD symptoms
	Tahir [64]	prospective, open-label	100	Turkey	7-repeat transmission more likely in MPH responders than non-responders
	Seeger [66]	prospective, open-label	47	Germany	decreased effects with 7-repeat allele in combination with serotonin transporter promoter polymorphism LL homozygosity
	Hamarman [65]	prospective, open-label	45	USA (New York)	7-repeat associated with need for higher MPH doses
	Van der Meulen [52]	retrospective report	82	The Nether-lands	borderline significant ($p = 0.09$) association between 7-repeat and increased effects
	Cheon [67]	prospective, open-label	83	Korea	increased effects on ADHD symptoms for 4-repeat homozygotes
	Zeni [53]	prospective, open-label	111	Brazil	no effect of the 7-repeat allele on ADHD symptoms
	Keresztúri [49]	prospective, open-label	122	Hungary	no effect of the 7-repeat allele on ADHD symptoms
	Tharoor [51]	retrospective report	159	USA (Missouri)	no effect of the 7-repeat allele on ADHD symptoms
Dopamine receptor (DRD5)	Tahir [64]	prospective, open-label	100	Turkey	151-bp allele transmission more likely in MPH responders than non-responders
Nicotinic acetyl-choline alpha4-receptor (CHRNA4)	Tharoor [51]	retrospective report	159	USA (Missouri)	no effect on ADHD symptoms
Norepinephrine transporter protein 1 (SLC6A2)	McGough [42]	prospective, double-blind, placebo-controlled, dose response	82	USA (California)	no effect of the exon 9 polymorphism on ADHD symptoms
	Yang [92]	prospective, open-label	45	China	decreased effects for homozygous A-allele of the exon 9 polymorphism

Table 1 (continued)

Gene	First author	Design	Sample size	Study location	Outcome
Serotonin receptors (HTR1B, HTR2A)	Zeni [53]	prospective, open-label	111	Brazil	no effect on ADHD symptoms
Serotonin transporter (5-HTT)	McGough [42]	prospective, double-blind, placebo-controlled, dose response	82	USA (California)	decreased effects on ADHD symptoms for those lacking the 12-repeat intron 2 polymorphism; improved math test performance on higher MPH doses for those lacking the promoter polymorphism L allele
	Seeger [66]	prospective, open-label	47	Germany	decreased effects for promoter polymorphism LL homozygotes in combination with DRD4 7-repeat allele
	Zeni [53]	prospective, open-label	111	Brazil	no effect of the promoter polymorphism on ADHD symptoms
	Tharoor [51]	retrospective report	159	USA (Missouri)	no effect of the promoter polymorphism on ADHD symptoms
Synaptosomal-associated protein (SNAP25)	McGough [41]	prospective, double-blind, placebo controlled	81	USA (6 sites)	T1069C C polymorphism associated with effects on teacher symptom ratings but not parent-teacher composite ratings
	McGough [42]	prospective, double-blind, placebo-controlled, dose response	82	USA (California)	no effect on ADHD symptoms

Studies are listed in chronologic order, with prospective double-blind placebo-controlled studies listed before naturalistic studies.

Table 2. Pharmacogenetic studies of methylphenidate effects on ADHD symptoms in adults

Gene	Study	Design	Sample size	Study location	Outcome
Dopamine transporter (SLC6A3)	Mick [55]	prospective, double-blind, placebo-controlled	66	USA (Boston)	no effect on ADHD symptoms
	Kooij [56]	prospective, double-blind, placebo-controlled	42	The Netherlands	decreased effects with homozygous 10-repeat
Dopamine receptor D4 (DRD4)	Kooij [56]	prospective, double-blind, placebo-controlled	42	The Netherlands	no effect on ADHD symptoms
Norepinephrine transporter (SLC6A2)	Kooij [56]	prospective, double-blind, placebo-controlled	42	The Netherlands	no effect of the NET promoter polymorphism on ADHD symptoms

mine [34, 35]. Cook et al. [36] first reported an association between ADHD and the 10-repeat (480-bp) allele of a variable number tandem repeat (VNTR) in the 3′-untranslated region (3′-UTR) of SLC6A3 which has been replicated in numerous, but not all, studies [12]. Moreover, several neuroimaging studies indicate that ADHD is associated with increased SLC6A3 densities in striatal regions [37], and that individuals with the 10-repeat allele exhibit 50% greater densities than other genotypes [38]. This suggests that ADHD medications which decrease dopamine reuptake might serve to attenuate the effects of underlying brain pathophysiology. Similarly, it has been hypothesized that functional SLC6A3 polymorphisms may influence response to ADHD pharmacotherapy. Below we summarize results from both prospective controlled studies and naturalistic SLC6A3 ADHD pharmacogenetic studies.

Placebo-Controlled SLC6A3 Methylphenidate Studies in Children

Four prospective double-blind placebo-controlled trials of SLC6A3 3′-UTR polymorphisms and their association with MPH effects on ADHD symptoms have been conducted in pediatric samples [39–42]. Stein et al. [39] found that the presence of one or two 10-repeat alleles was associated with higher rates of parent-rated symptom reduction and reduced impairment in 47 children treated with 18, 36, and 54 mg of OROS MPH. Individuals homozygous for the 9-repeat allele demonstrated a non-linear dose-response curve, had more stimulant-related side effects, and remained more impaired during treatment. Similar findings were reported for the 9/9 genotype group in a double-blind placebo-controlled trial of 159 Canadian children with ADHD conducted by Joober et al. [40]. Although 10-repeat carriers displayed a significant positive response to 10 mg MPH, 9/9 homozygotes displayed a negative response on parent but not teacher ratings [40]. In a study of 81 preschoolers with ADHD treated with MPH (Preschool ADHD Treatment Study; PATS), there were no significant effects of SLC6A3 polymorphisms on a composite measure of parent and teacher ratings [41]. However, on parent ratings of ADHD symptoms, there was a negative effect for 9/9 homozygotes. Finally, in a recent study of American school-age children, McGough et al. [42] found no significant effects of the 10-repeat allele on MPH efficacy or side effects; their method of defining genotype groupings precluded analysis of homozygous 9-repeat effects. In summary, 2 of the 4 placebo-controlled pediatric SLC6A3 trials found an improved MPH response for 10-repeat homozygotes, while 9-repeat homozygosity was associated with a diminished parent-rated medication response in all 3 studies that evaluated its effects.

As in Stein et al. [39] summarized above, 2 additional placebo-controlled reports suggest a relationship between SLC6A3 3′-UTR 9-repeat homozygosity and MPH side effects in children with ADHD [43, 44]. In their study of 177 participants with ADHD derived from 2 previously reported studies, Gruber et al. [43] evaluated associations between SLC6A3 polymorphisms and three MPH side effect factors: emotionality, somatic complaints, and over-focused. The study documented a significant association between 9-repeat homozygosity and increased scores on the emotionality factor (ir-

1

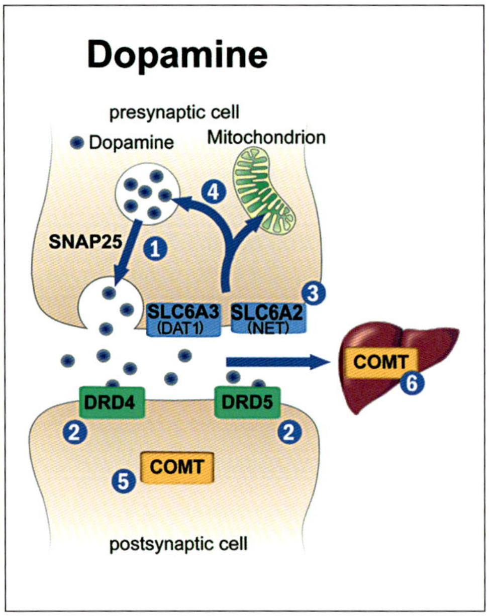

2

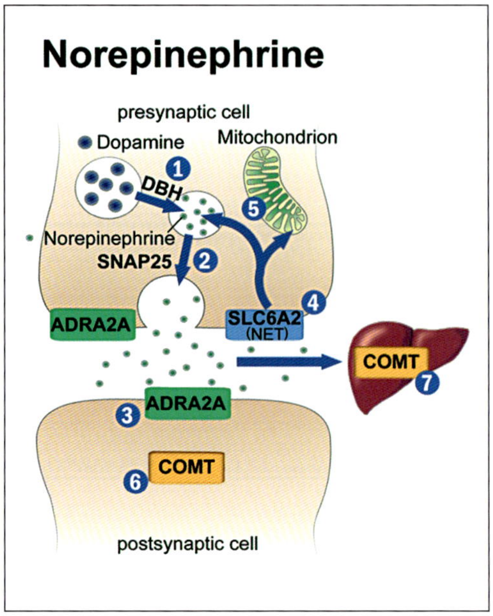

Fig. 1. Functions of dopamine-associated genes [124–128] linked to ADHD susceptibility and/or methylphenidate response. (1) Dopamine in storage vesicles is released into the synaptic cleft by synaptosomal-associated protein 25 (SNAP25)-mediated exocytosis. (2) Dopamine activates receptors on the presynaptic and postsynaptic cells, including dopamine receptor D4 (DRD4) and dopamine receptor D5 (DRD5). (3) The action of dopamine in the synaptic cleft is largely terminated via reuptake into the presynaptic cell by the dopamine transporter (SLC6A3). In areas of the brain such as the prefrontal cortex where SLC6A3 is not abundant, its function is thought to be assumed by the norepinephrine transporter (SLC6A2). (4) Some dopamine that re-enters the presynaptic cell is recycled in storage vesicles for later release, and some is transported into mitochondria to be broken down by monoamine oxidase. (5) Some of the dopamine released into the synaptic cleft is transported into postsynaptic cells and then catabolized by catechol-O-methyltransferase (COMT). (6) Dopamine remaining in the synaptic cleft diffuses into the bloodstream. It is inactivated in the liver by COMT and monoamine oxidase.

Fig. 2. Functions of norepinephrine-associated genes [124, 126, 129, 130] linked to ADHD susceptibility and/or methylphenidate response. (1) Dopamine in storage vesicles in the presynaptic neuron is converted to norepinephrine by dopa-β-hydroyxylase (DBH). (2) Norepinephrine in storage vesicles is released into the synaptic cleft by synaptosomal-associated protein 25 (SNAP25)-mediated exocytosis. (3) Norepinephrine activates receptors on the presynaptic and postsynaptic cells, including adrenergic α2A (ADRA2A). (4) The action of norepinephrine in the synaptic cleft is largely terminated by reuptake into the presynaptic cell by the norepinephrine transporter (SLC6A2). (5) Some norepinephrine that re-enters the presynaptic cell is recycled in storage vesicles for later release, and some is transported into mitochondria to be broken down by monoamine oxidase. (6) Some of the norepinephrine released into the synaptic cleft is transported into postsynaptic cells and then catabolized by catechol-O-methyltransferase (COMT). (7) Norepinephrine remaining in the synaptic cleft diffuses into the bloodstream. It is inactivated in the liver by COMT and monoamine oxidase.

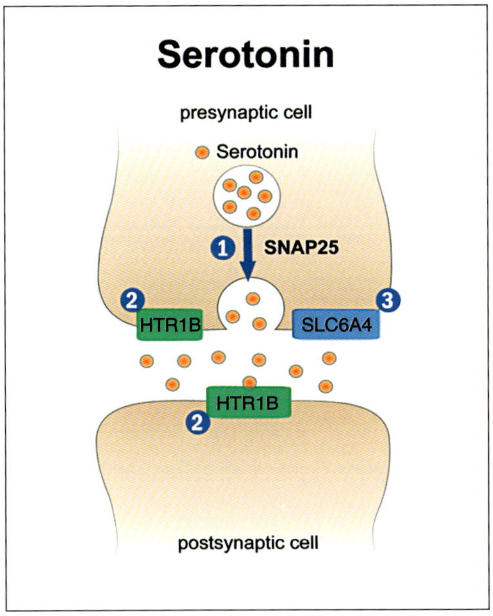

Fig. 3. Functions of serotonin-associated genes [124, 126, 131] linked to ADHD susceptibility and/or methylphenidate response. (1) Serotonin in storage vesicles is released into the synaptic cleft by synaptosomal-associated protein 25 (SNAP25)-mediated exocytosis. (2) Serotonin activates receptors on the presynaptic and postsynaptic cells, including the serotonin receptor HTR1B. (3) The action of serotonin in the synaptic cleft is largely terminated via reuptake into the presynaptic cell by the serotonin transporter (SLC6A4).

ritability, sadness, prone to cry, anxious) that worsened with MPH treatment, while children with 10-repeat homozygosity had significant increases in the somatic complaints factor (decreased appetite, stomachache, headache, insomnia) during MPH treatment [43]. In another study, Leddy et al. [44] investigated MPH effects on eating in 58 children who participated in an ADHD therapeutic summer camp, and also documented significant associations between 9-repeat homozygosity and MPH side effects, although the findings differ from those of Gruber at al. [43]. While Leddy et al. [44] did not find significant relationships between SLC6A3 genotype and ratings of stomachache or loss of appetite with MPH treatment, they did show a significantly greater suppression of lunch intake as MPH dose increased for 9-repeat homozygotes.

Naturalistic SLC6A3 Methylphenidate Studies in Children

The 11 naturalistic pharmacogenetic studies of SLC6A3 3′-UTR-polymorphisms in children with ADHD have not yielded consistent results. For example, only 1 of the naturalistic pediatric trials identified a link between the SLC6A3 10-repeat allele and improved MPH response. This analysis, based on parental retrospective report in 119

Irish children, found that 10-repeat carriers were more likely to have an improved MPH response [45]. In this study, a linear relationship existed between the number of 10-repeats and degree of improvement.

In contrast, the homozygous SLC6A3 10-repeat allele has been associated with diminished stimulant response in 4 pediatric naturalistic studies [38, 46–48]. In their study of 30 stimulant-naïve African-American youths with ADHD, Winsburg and Cumings reported that 86% of non-responders were 10-repeat homozygotes compared with 31% of responders [38]. In a sample of 50 European-Brazilian males with ADHD who underwent open titration with MPH up to 0.7 mg/kg/day, individuals who failed to show a ≥50% reduction in baseline ADHD ratings with MPH treatment were significantly more likely to be 10-repeat homozygotes [47]. A third study in 11 Korean subjects found that only 27% of 10-repeat homozygotes met MPH response criteria compared with 100% of subjects without this genotype [46]. Similarly, a recent study of 141 French children found that 10-repeat homozygotes were overrepresented in the low MPH response group, as defined by having a less than 2-point improvement in CGI-Severity score [48].

No significant effect of the SLC6A3 10-repeat allele on MPH response was documented in the remaining 6 uncontrolled studies of children with ADHD [49–54]. These included prospective open-label samples of 82 Dutch children (treated with <0.6 mg/kg/day) [52], 122 Hungarian children (mean dose 0.55 mg/kg/day) [49], 26 Irish children (mean dose 0.6 mg/kg/day) [50], and 111 Brazilian youth (mean dose 0.5 mg/kg/day) [53], as well as 2 retrospective analyses examining 168 youth in the UK [54] and 156 children from the USA [51] (mean medication doses not specified).

Meta-Analysis of SLC6A3 Methylphenidate Studies in Children

Purper-Ouakil et al. [48] conducted a meta-analysis of SLC6A3 10-repeat allele effects on MPH responder versus non-responder status, combining the results of 6 studies for a total of 475 children. This meta-analysis, which evaluated 5 naturalistic studies and 1 prospective controlled trial, found that 10-repeat homozygotes were less likely to show a moderate to good MPH response compared to other genotypes.

SLC6A3 Methylphenidate Studies in Adults

To date, two SLC6A3 pharmacogenetic studies – both of which are placebo-controlled double-blind trials – have been conducted in adult samples [55, 56]. The first reported no link between SLC6A3 3′-UTR genotype and response in 66 subjects titrated to a maximum MPH dose of 1.3 mg/kg/day [55]. However, the sample included only three 9-repeat homozygotes, limiting statistical power to detect an effect for this group. The second study, which titrated MPH doses to a maximum of 1.0 mg/kg/day, found that carriers of a single SLC6A3 10-repeat allele were more likely to have a favorable medication response compared to 10-repeat homozygotes [56]. Of note, the single 9/9 homozygous patient was not included in the analysis. Thus, it is unclear if adults with the 9/9 genotype differ from other groups in their MPH response.

SLC6A3 Haplotype Methylphenidate Studies

There has been increasing interest in the effects of a SLC6A3 haplotype involving the 3′-UTR 10-repeat allele and a 3-repeat allele of 30-bp VNTR in intron 8 (SLC6A3 10/3). Intriguingly, 3 studies of the SLC6A3 10/3 haplotype have documented an association with ADHD [57–59]. However, the first pediatric pharmacogenetic study of the SLC6A3 10/3 haplotype failed to document significant effects of SLC6A3 10/3 on MPH response [50]. Study limitations included the relatively small sample size (n = 26) and the naturalistic study design.

SLC6A3 Amphetamine Studies

As yet, there are no reported SLC6A3 pharmacogenetic studies of amphetamine (AMPH) in ADHD. However, Lott et al. [32] examined the relationship between SLC6A3 and AMPH response in healthy college students in a placebo-controlled crossover trial. In this study, 9-repeat homozygotes were less able to 'feel' AMPH effects relative to other groups. This finding is consistent with prior suggestions of a differential effect of stimulants on 9-repeat homozygotes compared to 10-repeat carriers [39–41].

Dopamine Receptor D4

DRD4 encodes a dopamine receptor that regulates dopamine synthesis and release, as well as the firing rate of dopamine neurons [60]. The association of a 7-repeat (48-bp) VNTR polymorphism in DRD4's exon 3 with ADHD is one of the most replicated findings in psychiatric genetics [12]. In vitro studies indicate that the 7-repeat variant is less responsive to dopamine [61, 62], suggesting a functional role for this polymorphism in stimulant medication response. In addition, a 240-bp polymorphism in the DRD4 promoter has been linked to ADHD etiology [63], and has also piqued the interest of ADHD pharmacogenetic investigators [41, 42].

Placebo-Controlled DRD4 Methylphenidate Studies

Thus far, only one placebo-controlled DRD4 pharmacogenetic trial in school-age children has been reported. In a study of 82 children with ADHD, McGough et al. [42] found no significant associations between the DRD4 exon 3 or promoter polymorphisms in terms of MPH effects on ADHD symptoms, but did find significant MPH dose × gene interactions for math test performance. Those without the 4-repeat exon 3 genotype (–4/–4) had a deterioration in math performance at higher MPH doses, while those lacking the 240-bp promoter (–240/–240) polymorphism had improved math performance at increased doses. Further, the investigators evaluated associations between DRD4 polymorphisms and four MPH side effect factors: irritability, vegetative symptoms, abnormal movements, and somatic symptoms. The study showed that the 240-bp promoter polymorphism homozygotes (+240/+240) had decreased irritability (picking, worried/anxious, crabby/irritable, and tearful/sad/depressed) and somatic symptoms (headache and stomachache) with MPH treatment.

Additional prospective double-blind placebo-controlled trials have been conducted in adults [56] and in preschoolers with ADHD [41]. In a study of adults with ADHD, Kooij et al. [56] found no association between the DRD4 exon 3 or promoter polymorphisms and MPH response in their study of 42 adults with ADHD. Similarly, the PATS trial did not identify significant effects for DRD4 exon 3 polymorphisms in preschool children in terms of symptom reduction or dose-response effects based upon parent, teacher, or parent-teacher composite ratings, although associations were seen with several MPH side effects [41]. The presence of the 4-repeat allele predicted the side effect of abnormal picking, while the 7-repeat allele was associated with social withdrawal as MPH dose increased. In addition, the PATS trial described an association between DRD4 promoter +240/+240 homozygotes and improved MPH response in terms of ADHD symptoms, although +240/+240 homozygotes were also more likely to become crabby or irritable with increasing dose [41].

Naturalistic DRD4 Methylphenidate Studies

Findings in naturalistic DRD4 MPH pharmacogenetic studies have been mixed, as 4 of the 9 naturalistic studies found no significant link between MPH response and the 7-repeat allele, including a retrospective study of 159 US children [51], and prospective open-label studies of 30 US children [38], 111 Brazilian children [53], and 112 Hungarian children [49]. However, a prospective open-label study of 100 Turkish children showed that transmission of the 7-repeat allele is more likely in MPH responders compared to non-responders [64], and a retrospective study of 82 Dutch children found a borderline significant association between the 7-repeat allele and better MPH response [52]. In contrast, 2 prospective open-label studies suggest an association between the DRD4 7-repeat allele and diminished MPH response [65–67]. This is consistent with prior indications that this variant encodes a dopamine receptor that is hyporesponsive to its agonist [61, 62]. Specifically, Hamarman et al. [65] demonstrated in 45 subjects that 7-repeat carriers required higher MPH doses for optimal symptom reduction. In a study of 47 German subjects, Seeger et al. [66] found that children with at least one DRD4 7-repeat allele and homozygosity for the long allele of a serotonin transporter promoter polymorphism had less improvement in functioning with MPH treatment compared to those with other genotype combinations. In addition, in a study of 83 Korean children, DRD4 4-repeat homozygotes were more likely to exhibit positive MPH responses on parent and teacher ratings than those with other genotypes [67], but it should be noted that the 7-repeat genotype is extremely rare in Asian samples.

DRD4 Atomoxetine Studies

One study found that children carrying the DRD4 4-repeat allele showed a trend towards improved response on atomoxetine [68]. In addition, improvement in the hyperactivity symptom domain was maximized in the absence of any 7-repeat variant [68].

Adrenergic α2A Receptor

ADRA2A encodes a norepinephrine autoreceptor whose activation dampens the cell firing rate and limits norepinephrine release [69]. Several animal studies suggest that α2 noradrenergic receptors may mediate MPH effects, with administration of α2 adrenoceptor antagonists blocking MPH beneficial effects [70, 71]. A –1291 C>G single-nucleotide polymorphism (SNP) creates an *Msp*I site in the ADRA2A promoter region [72] that may be both functional [73] and linked to inattention symptoms [22–24]. Moreover, effects of the C1291G polymorphism on ADHD symptom scores have been documented in 3 pharmacogenetic studies thus far [74–76]. The first assessed 104 children of mixed ADHD subtypes after MPH treatment [74], and reported that G allele carriers showed significantly improved MPH response on inattention scores, but not hyperactive-impulsive scores. A subsequent naturalistic study by da Silva et al. [75] is the first ADHD pharmacogenetic study of the ADHD-inattentive type. In this study, G allele carriers displayed significantly lower inattentive scores after MPH treatment compared to those lacking the G allele. Similarly, Cheon et al. [76] evaluated 114 Korean children with ADHD in a prospective open-label study, and found that G homozygotes had significantly greater rates of MPH 'good response', as well as greater improvements in total ADHD symptom scores, compared to other genotypes. However, in contrast to previous studies which noted specific effects on attentional symptoms, the findings of Cheon et al. [76] were not significant when the outcome was inattentive symptom scores as opposed to total symptom scores.

Carboxylesterase 1

MPH undergoes esterification in the blood stream via CES1 to D/L ritalinic acid and L-ethylphenidate [77]. Zhu et al. [78] described a mutation in exon 4 of CES1 at codon 143 leading to a nonconservative amino acid substitution (gly to glu) which is associated with complete loss of hydrolytic activity toward MPH. Recently, Nemoda et al. [79] investigated the effects of this polymorphism in 122 Hungarian children with ADHD in a prospective open-label study, and found that glu allele carriers required significantly lower MPH doses for symptom reduction compared to gly/gly homozygotes, presumably due to higher plasma drug levels for a given dose.

Catechol-O-Methyltransferase

COMT catabolizes dopamine and norepinephrine. COMT has a functional polymorphism at codon 158 that results in a single amino acid change (met for val). Enzyme activity for the variants are as follows: val/val homozygotes have high, val/met heterozygotes have intermediate, and met/met homozygotes have 4–5 times lower COMT activity [80]. Although some prior studies have identified associations between ADHD and COMT polymorphisms [20, 21], a meta-analysis of 13 studies did not find a significant association between the COMT val158met polymorphism and ADHD [81]. Nonetheless, COMT's role in catecholamine catabolism, combined with clear of evidence of the codon 158 polymorphism's functionality, makes it a compelling ADHD pharmacogenetics study candidate.

To date, 1 double-blind placebo-controlled trial and 2 prospective open-label studies have investigated the link between COMT polymorphisms and MPH response [42, 49, 82]. In a controlled trial, McGough et al. [42] found val/val homozygosity to be associated with diminished MPH effects on ADHD symptoms as well as increased irritability with MPH treatment. In contrast, the 2 open-label studies linked val/val homozygosity to improved MPH response [49, 82]. Kereszturi et al. [49] documented a significant interaction between the COMT val/val genotype and good MPH response in terms of hyperactive-impulsive but not inattentive symptoms in a sample of 122 Hungarian children with ADHD. Cheon et al. [82] also found a trend suggesting an association between COMT val/val homozygosity and improved MPH effects on ADHD symptoms, as well as a significant association between COMT met/met homozygosity and poor MPH response as rated by teachers but not parents.

COMT Amphetamine Studies

One prior study has examined the relationship between COMT polymorphisms and AMPH response. Mattay et al. [83] documented that working memory efficiency, assessed via functional MRI, was enhanced by AMPH administration for val/val subjects, while AMPH had adverse effects under high working memory load conditions for met/met subjects.

Dopamine Receptor D5

DRD5 is a G-protein coupled receptor that stimulates the production of adenylate cyclase. Studies of D5 null mice suggest that DRD5 may contribute to the activation of dopaminergic pathways relevant to exploratory locomotion and prepulse inhibition [84]. Three meta-analyses have documented a significant association between ADHD and the 148-bp allele of a microsatellite marker located 5′ to the DRD5 gene [85–87]. Review of the literature indicates a single DRD5 pharmacogenetic study. This study by Tahir et al. [64] in 100 Turkish children did not identify any children with the 148-bp allele, but did find an association between the 151-bp allele of the DRD5 5′ microsatellite marker and favorable MPH response.

Norepinephrine Transporter Protein 1 (SLC6A2)

SLC6A2 encodes a presynaptic protein involved in reuptake of norepinephrine from the synaptic cleft [88], as well as dopamine reuptake from the synapse in certain parts of the brain, including the prefrontal cortex [89] (fig. 2). Stimulant medications block reuptake at norepinephrine transporters [90], and norepinephrine transporter blockade is also the presumed mechanism of action for the non-stimulant ADHD medication atomoxetine [91]. Several SLC6A2 polymorphisms have been associated with ADHD [25, 26]. Thus far, 2 studies have evaluated the link between a G1278A polymorphism at exon 9 of SLC6A2 and MPH response. In a double-blind placebo-controlled trial, McGough et al. [42] did not find a link between the G1278A polymorphism and MPH response or side effects. In contrast, in an open-label study of 45 Han Chinese youths with ADHD, Yang et al. [92] found that A/A homozygotes had a diminished MPH response in terms of hyperactive-impulsive but not inattentive symptoms compared with other genotypes. However, since the G1278A allele has no known functional activity, Yang et al. [92] note that it might be in linkage disequilibrium with another allele responsible for outcome differences. Recently, an additional SLC6A2 pharmacogenetic study conducted in an adult sample evaluated the association between MPH response and a 4-bp insertion/deletion polymorphism in the promoter region of SLC6A2 [56]. This double-blind placebo-controlled study by Kooij et al. [56] did not identify a significant relationship between the SLC6A2 promoter polymorphism and medication response.

SLC6A2 Amphetamine Studies
The association between eight SLC6A2 polymorphisms and subjective response to D-AMPH was evaluated in a prospective placebo-controlled double-blind study of 99 healthy German adults [93]. This study found that the CC genotype of the 36001A/C SNP and the GCC haplotype from the 28257G/C, 28323C/T, and 36001A/C SNPs were linked to higher self-reported positive mood after AMPH administration. The authors note that although no functional consequences of the CC genotype or GCC haplotype are presently known, these polymorphisms are located in transcription-factor binding sites, and thus may alter SLC6A2 transcription rates [93].

SLC6A2 Atomoxetine Studies
Recently, Ramoz et al. [94] investigated whether 108 polymorphisms and 8 haplotype blocks in SLC6A2 influenced response to atomoxetine in 2 independent randomized double-blind trials of 160 and 105 children with ADHD. A haplotype block spanning exons 4–9 of SLC6A2, where 36 SNPs have been genotyped, was significantly associated with atomoxetine response in both independent cohorts and the combined cohort. In addition, significant associations between 20 different SLC6A2 SNPs and atomoxetine response were observed when analyses did not account for multiple comparisons; the authors note that no individual SNP reached statistical significance at the multiple comparison threshold.

Serotonin Transporter (SLC6A4)

SLC6A4 encodes a presynaptic protein responsible for serotonin reuptake from the synaptic cleft (fig. 3). Serotonin transporter polymorphisms have also been implicated in ADHD etiology, with the long (L) allele of a 44-bp insertion/deletion promoter polymorphism linked to hyperkinetic disorder [12, 95]. Of note, functionality has been shown for the promoter polymorphism, with L homozygosity yielding higher levels of transporter function than the L/S or S/S genotypes [96]. In addition, an SLC6A4 intron 2 VNTR polymorphism has been associated with transporter expression, with the 12-repeat allele linked to increased transporter transcription [97, 98].

Serotonin transporter polymorphisms have also been evaluated in pharmacogenetic investigations. Results for the promoter polymorphism have been mixed, with 2 studies showing significant effects on MPH response and 2 studies showing no effect. In a double-blind placebo-controlled trial, McGough et al. [42] found improved math test performance on higher MPH doses for those lacking the L allele, although significant effects on ADHD symptoms were not seen. In a prospective open-label trial, Seeger et al. [66] documented decreased effects of MPH for individuals with L homozygosity plus the DRD4 7-repeat allele. In contrast, the prospective open-label trial of Zeni et al. [53] and the retrospective report of Tharoor et al. [51] did not find significant effects of the SLC6A4 promoter polymorphism. Only one study thus far has examined the relationship between the intron 2 polymorphism and MPH effects. This double-blind placebo-controlled trial by McGough et al. [42] documented decreased MPH effects on ADHD symptoms for those lacking the 12-repeat intron 2 polymorphism.

SLC6A4 Amphetamine Studies

Lott et al. [99] evaluated the association between the SLC6A4 promoter and intron 2 polymorphisms and subjective response to D-AMPH in a double-blind placebo-controlled study of 101 healthy adults. The trial documented that intron 2 10-repeat homozygotes experienced significantly greater euphoria with AMPH administration compared to the other genotypes. In addition, although the finding was not statistically significant, there was some indication that participants homozygous for both the promoter polymorphism L allele and the intron 2 12-repeat allele had the weakest subjective AMPH response.

Synaptosomal-Associated Protein 25

SNAP25 is a vesicle docking protein involved in neurotransmitter exocytosis from storage vesicles into the synaptic space [100]. Several studies have examined the link between ADHD and two SNPs (T1069C and T1065G) at the 3′ end of SNAP25 [101–

104]. Although study results have not been entirely consistent, pooled analyses for T1065G reveal significant evidence of an association with ADHD [12]. Studies of the mouse mutant strain *coloboma,* which has a SNAP25 deletion [105], suggest a role for SNAP25 in MPH response. Specifically, hyperactivity in the *coloboma* mouse was suppressed by AMPH but not MPH administration [105]. This is consistent with presumed differences in these stimulants' mechanisms of action, since AMPH, but not MPH, compensates for reduced exocytotic catecholamine release by reversing the catecholamine diffusion gradient across the dopamine transporter.

Thus far, 2 double-blind placebo-controlled trials have evaluated the effects of SNAP25 polymorphisms on MPH response. In a study of school-age children, McGough et al. [42] did not find significant associations between SNAP25 polymorphisms and MPH efficacy or side effects. However, the PATS trial of preschool children with ADHD found that T1065G T homozygotes had moderately improved MPH dose responses, while T1069C T homozygotes exhibited poorer MPH responses [42]. In addition, T1065G G homozygotes were more likely to develop sleep difficulties and irritability than T allele carriers, while T1069C C homozygotes were more likely to develop tics and other abnormal movements compared to T allele carriers.

Amphetamine and Atomoxetine Metabolic Pathways

Although effects on drug metabolism and pharmacokinetics frequently provide the basis for pharmacogenetic investigations [106], ADHD pharmacogenetic studies have primarily investigated the effects of genetic variability on drug targets, such as transporters and receptors [13, 27]. To date, the literature contains a single study documenting the effects of a MPH metabolism polymorphism (the CES1 143glu-variant) on ADHD treatment and dose response [79]. Genetic effects on AMPH and atomoxetine metabolism are also of interest.

Amphetamine Metabolism

AMPH undergoes metabolism via hepatic cytochrome P450 (CYP) isozymes. AMPH is metabolized along 2 major pathways – CYP 2D6 and CYP 3A4 – which are differentially employed by various species [107]. In the first pathway, hydroxylation of AMPH via CYP 2D6 yields *p*-hydroxy-AMPH. Although CYP 2D6 is believed to play only a minor role in human AMPH metabolism [108], up to 20% of Caucasians are poor metabolizers, which could have implications for dosing and medication tolerability in individual patients [109].

In the second pathway, which is dominant in humans, AMPH undergoes deamination via CYP 3A4 to L-phenyl-propane-2-one. In a study of mixed AMPH salts, mean drug plasma concentrations following acute dosing were 25% higher in African-American children [48]. Intriguingly, previous studies have reported racial differences in CYP-3A4-mediated drug metabolism, with Caucasian subjects demonstrating the highest levels of activity, and 1 allelic variant that is heterozygous in 64% of African-Americans associated with decreased metabolic activity [110]. Although a definitive association between CYP 3A4 polymorphisms and racial differences in AMPH metabolism has not been demonstrated, this may represent a promising area for future study.

Atomoxetine Metabolism
Atomoxetine is metabolized by the CYP 2D6 isozyme system, and subjects' CYP 2D6 status influenced dosing titration algorithms and subsequently derived approved dosing limits during atomoxetine drug development. A recent meta-analysis of atomoxetine studies found that poor CYP 2D6 metabolizers displayed greater symptom improvement than extensive metabolizers, presumably due to higher plasma drug levels, and were more likely to remain in therapy [111]. However, diminished appetite and increased tremor were reported in poor CYP 2D6 metabolizers, who also had greater medication-related changes in pulse and blood pressure [111].

Genome-Wide Approaches

In contrast to candidate gene studies which presume some knowledge of the underlying biological system and require specific hypotheses regarding the polymorphisms under investigation, genome-wide studies do not require a priori hypotheses related to specific genes but scan the entire genome to identify 'hot spots' related to outcome. To date, 2 genome-wide ADHD treatment investigations have been conducted. One prior study employed quantitative trait analysis in a genome-wide scan assessing for linkage with MPH response [112]. A linkage peak of moderate significance was found on chromosome 7, with additional peaks on chromosomes 3, 5, and 9. Further study, including genome-wide association with high density SNP chips, will be necessary to identify the specific genes corresponding to the observed linkage peaks. An additional genome-wide association study recently evaluated response to a MPH transdermal system [113]. In this open-label study of 187 children with ADHD, the strongest association ($p = 3 \times 10^{-6}$) fell short of the threshold for statistical significance in a genome-wide association study. However, intriguing non-significant associations were suggested in the glutamate receptor 7 gene and in two SNPs within the norepinephrine transporter gene.

Current Research Challenges and Future Directions

While the results of prior ADHD pharmacogenetic and pharmacogenomic studies have been intriguing, many challenges must still be addressed. For example, the majority of ADHD pharmacogenetic studies published to date have examined short-term response to MPH. Although MPH is a frequent first-line treatment, it is important to study genetic predictors of response in additional ADHD medications, including AMPH preparations and atomoxetine.

Study design differences may partially account for the heterogeneity of current findings. For example, in the case of the dopamine transporter studies, more consistent findings appear to be emerging from placebo-controlled studies of children with ADHD which evaluate a range of doses. Open-label or retrospective assessment in which medication doses are not specified or are lower than those used in efficacy studies might bias against finding significant treatment effects [27]. Moreover, when studies are not placebo controlled it may be more difficult to detect a pharmacogenetic effect since placebo responders are not differentiated from active drug responders.

Pharmacogenetic studies are also constrained by the type of outcome measures used, as many studies rely on simple dichotomous outcomes (e.g. responder versus non-responder), which have limited power to detect effects compared with quantitative measures. Correlations between multiple outcome measures in the same subjects are also known to be fairly weak, raising the question as to which measure best defines positive response [114]. In several cases, study results have differed depending on whether parents, teachers, or composites are used to define outcomes [40, 41]. Consequently, investigators should be cautious in prematurely selecting a single outcome measure or combination. Until more is learned about defining the phenotype of medication response, multiple methods should be used.

An additional methodological issue is how genotypes are grouped for analysis. In order to minimize the potential for spurious findings and type I errors, investigators must limit their analyses to minimal genotype combinations. For some genes, the ADHD risk polymorphisms are the less common variants (e.g. DRD4 7-repeat allele), while for other genes (e.g. dopamine transporter SLC6A3), it is the more common variants that are associated with the disorder. For SLC6A3, the 10/10 and 10/9 genotypes are most common, and earlier studies combined these common genotypes. This practice assumed a dominant effect of either the SLC6A3 9 or 10 allele, but failed to test for a recessive effect of the 9/9 genotype. Alternative grouping of genotypes based on the presence of one or more SLC6A3 9-repeat alleles has led to different results. Future candidate gene studies would benefit from consensus on optimal strategies to define genotype groupings. Genotypes should not be lumped together when evidence of one allele's dominance is lacking in previous pharmacogenetic studies.

Variation in sample size and composition may also contribute to differences in study results. Modest sample sizes have limited statistical power to detect mild or moderate genetic effects. Another potential contributor to the observed discrepancies in study findings is that pharmacogenetic effects may vary in different ethnic and racial groups. This suggests that the genetic variants being studied may not be causing the effect observed, but instead may be in linkage disequilibrium with the actual functional genetic variants. Furthermore, previous investigations may have failed to identify consistent genetic effects due to differences in sample subtype composition, given evidence in some prior studies that certain genetic variants may influence response to medication in terms of hyperactive-impulsive symptoms [49] or inattentive symptoms [74] but not both domains.

In addition, although ADHD pharmacogenetic studies to date have not evaluated interactions with additional environmental exposures, prior evidence hints that such exposures may be important modifiers of genetic effects on medication response. For example, tobacco smoke exposure may influence dopamine release by interacting with both catecholamine-related genetic variants [115, 116] and MPH [117]. If the suggested 3-way interaction between catecholamine genes, MPH, and tobacco exposure is in fact at play, we would expect measured associations between genotype and medication response to vary according to the tobacco exposure level of the different study populations.

Furthermore, failure to evaluate gene-gene interactions, which have received little attention in ADHD pharmacogenetic studies to date [66], may also be obscuring effects. In addition, it is increasingly recognized that drug response is the result of a complex matrix of factors, rather than a single factor [30]. As a result, experts have proposed that future pharmacogenetic studies shift their focus from individual genes to pathways encompassing genes for drug-metabolizing enzymes and transporters, as well as genes encoding drug targets and their downstream signals [118].

Potential Clinical Applications

Despite great hopes, the potential of candidate gene association studies to yield clinically relevant information regarding ADHD medication response has not yet been realized. Concerns have been raised because the effects of common polymorphisms on drug response have typically been small. Hence, knowledge of small effects due to single polymorphisms may be of dubious clinical utility given the large effect sizes attributed to ADHD stimulant medications in general [5]. Moreover, drug response is increasingly recognized to be the result of a multitude of factors, rather than variations in a single gene. Ultimately then, pharmacogenetic study of individual candidate polymorphisms may not provide the tools for definitive determination of ADHD medication response, but rather may contribute to the development of clinically salient treatment prediction algorithms that incorporate a complex interplay of genetic and environmental factors.

Prediction of side effect risk and medication tolerability may be a practical future application for ADHD pharmacogenetic data. Stimulant medications are the recommended first-line treatments for ADHD [119]. However, open-label follow-up studies have shown that fewer than 60% of previously stabilized patients remained on stimulants after 12 months of treatment, although those who continued treatment showed sustained improvements from baseline [120, 121]. In one 5-year prospective investigation documenting the discontinuation of ADHD medication by the second study year in over half of participants, the authors concluded that side effects were major factors in patients' decisions to discontinue treatment [122]. In the PATS trial, development of irritability and increased emotionality were 2 major reasons subjects discontinued medication therapy [123]. Interestingly, pharmacogenetic analyses in the PATS trial and several other studies have revealed genetic predictors of numerous stimulant side effects, including irritability, emotionality, and somatic complaints [39, 41, 43, 44]. Conceivably then, awareness of increased side effect risk derived from pharmacogenetic data could be used to steer individuals toward tailored treatment regimens that are more likely to be tolerated over time.

The development of novel treatments may also provide an important clinical application for ADHD pharmacogenetics and pharmacogenomics findings. Further knowledge of genes that predict ADHD treatment response may aid in the development of more specific and efficacious medications for subsets of children with ADHD. Ultimately, it is hoped that pharmacogenetic and pharmacogenomic research will allow clinicians to tailor individual treatment choices based on genotype.

Conclusion

ADHD pharmacogenetics and pharmacogenomics research efforts are expanding worldwide. To date, several promising findings related to prediction of symptom response and side effects have been reported, although results have not been entirely consistent. Upcoming investigations should employ more standardized study designs while examining a wider range of stimulant and non-stimulant medications and a variety of outcome measures and informants. Fruitful avenues for future ADHD pharmacogenetic investigation may include study of polymorphisms in drug-metabolizing enzymes, as well as approaches that incorporate gene-gene interactions and effect modification by additional environmental exposures. Furthermore, investigators are increasingly interested in going beyond the study of single candidate genes to explore whole-genome approaches. Further research, likely involving multi-site collaborations to obtain larger samples, is clearly necessary before preliminary findings can be applied to clinical practice. Nonetheless, the promise of ADHD pharmacogenetics and pharmacogenomics is far reaching, and includes the potential to develop individualized medication regimens that improve symptom response, lessen risk of side effects, and increase long-term tolerability.

Acknowledgments

This work was supported, in part, by grant No. MH070564 (M.A.S.) and grant No. K23MH083881 (T.F.) from the National Institute of Mental Health.

References

1 Diagnostic and Statistical Manual of Mental Disorders: DSM-IV, ed 4. Washington, American Psychiatric Association, 1994.

2 Swanson J, Castellanos FX, Murias M, LaHoste G, Kennedy J: Cognitive neuroscience of attention deficit hyperactivity disorder and hyperkinetic disorder. Curr Opin Neurobiol 1998;8:263–271.

3 Barkley RA: Major life activity and health outcomes associated with attention-deficit/hyperactivity disorder. J Clin Psychiatry 2002;63(suppl 12):10–15.

4 Barbaresi WJ, Katusic SK, Colligan RC, Weaver AL, Leibson CL, Jacobsen SJ: Long-term stimulant medication treatment of attention-deficit/hyperactivity disorder: results from a population-based study. J Dev Behav Pediatr 2006;27:1–10.

5 Faraone SV, Biederman J, Spencer TJ, Aleardi M: Comparing the efficacy of medications for ADHD using meta-analysis. MedGenMed 2006;8:4.

6 Thiruchelvam D, Charach A, Schachar RJ: Moderators and mediators of long-term adherence to stimulant treatment in children with ADHD. J Am Acad Child Adolesc Psychiatry 2001;40:922–928.

7 Lowe N, Barry E, Gill M, Hawi Z: An overview of the pharmacogenetics and molecular genetics of ADHD. Current Pharmacogen 2006;4:231–243.

8 Charach A, Ickowicz A, Schachar R: Stimulant treatment over five years: adherence, effectiveness, and adverse effects. J Am Acad Child Adolesc Psychiatry 2004;43:559–567.

9 Clinical practice guideline: diagnosis and evaluation of the child with attention-deficit/hyperactivity disorder. American Academy of Pediatrics. Pediatrics 2000;105:1158–1170.

10 Faraone SV, Biederman J: Neurobiology of attention-deficit hyperactivity disorder. Biol Psychiatry 1998;44:951–958.

11 Volkow ND, Wang GJ, Fowler JS, Ding YS: Imaging the effects of methylphenidate on brain dopamine: new model on its therapeutic actions for attention-deficit/hyperactivity disorder. Biol Psychiatry 2005;57:1410–1415.

12 Faraone SV, Perlis RH, Doyle AE, Smoller JW, Goralnick JJ, Holmgren MA, Sklar P: Molecular genetics of attention-deficit/hyperactivity disorder. Biol Psychiatry 2005;57:1313–1323.

13 Polanczyk G, Zeni C, Genro JP, Roman T, Hutz MH, Rohde LA: Attention-deficit/hyperactivity disorder: advancing on pharmacogenomics. Pharmacogenomics 2005;6:225–234.

14 Weber WW: Pharmacogenetics. New York, Oxford Press, 1997.

15 Phillips KA, Veenstra DL, Sadee W: Implications of the genetics revolution for health services research: pharmacogenomics and improvements in drug therapy. Health Serv Res 2000;35:128–140.

16 Aitchison KJ, Gill M: Pharmacogenetics in the postgenomic era; in Plomin R, Devries J, Craig I, McGuffin P (eds): Behavioral Genetics in the Postgenomic Era. Washington, American Psychological Association, 2003.

17 Evans WE, Relling MV: Pharmacogenomics: translating functional genomics into rational therapeutics. Science 1999;286:487–491.

18 Solanto MV: Neuropsychopharmacological mechanisms of stimulant drug action in attention-deficit hyperactivity disorder: a review and integration. Behav Brain Res 1998;94:127–152.

19 Albayrak O, Friedel S, Schimmelmann BG, Hinney A, Hebebrand J: Genetic aspects in attention-deficit/hyperactivity disorder. J Neural Transm 2008; 115:305–315.

20 Eisenberg J, Mei-Tal G, Steinberg A, Tartakovsky E, Zohar A, Gritsenko I, Nemanov L, Ebstein RP: Haplotype relative risk study of catechol-O-methyltransferase (COMT) and attention deficit hyperactivity disorder (ADHD): association of the high-enzyme activity val allele with ADHD impulsive-hyperactive phenotype. Am J Med Genet 1999;88:497–502.

21 Qian Q, Wang Y, Zhou R, Li J, Wang B, Glatt S, Faraone SV: Family-based and case-control association studies of catechol-O-methyltransferase in attention deficit hyperactivity disorder suggest genetic sexual dimorphism. Am J Med Genet B Neuropsychiatr Genet 2003;118:103–109.

22 Roman T, Schmitz M, Polanczyk GV, Eizirik M, Rohde LA, Hutz MH: Is the alpha-2A adrenergic receptor gene (ADRA2A) associated with attention-deficit/hyperactivity disorder? Am J Med Genet B Neuropsychiatr Genet 2003;120:116–120.

23 Roman T, Polanczyk GV, Zeni C, Genro JP, Rohde LA, Hutz MH: Further evidence of the involvement of alpha-2A-adrenergic receptor gene (ADRA2A) in inattentive dimensional scores of attention-deficit/hyperactivity disorder. Mol Psychiatry 2006;11: 8–10.

24 Schmitz M, Denardin D, Silva TL, Pianca T, Roman T, Hutz MH, Faraone SV, Rohde LA: Association between alpha-2A-adrenergic receptor gene and ADHD inattentive type. Biol Psychiatry 2006;60: 1028–1033.

25 Bobb AJ, Addington AM, Sidransky E, Gornick MC, Lerch JP, Greenstein DK, Clasen LS, Sharp WS, Inoff-Germain G, Wavrant-De Vrieze F, Arcos-Burgos M, Straub RE, Hardy JA, Castellanos FX, Rapoport JL: Support for association between ADHD and two candidate genes: NET1 and DRD1. Am J Med Genet B Neuropsychiatr Genet 2005;134:67–72.

26 Kim CH, Hahn MK, Joung Y, Anderson SL, Steele AH, Mazei-Robinson MS, Gizer I, Teicher MH, Cohen BM, Robertson D, Waldman ID, Blakely RD, Kim KS: A polymorphism in the norepinephrine transporter gene alters promoter activity and is associated with attention-deficit hyperactivity disorder. Proc Natl Acad Sci USA 2006;103:19164–19169.
27 McGough JJ: Attention-deficit/hyperactivity disorder pharmacogenomics. Biol Psychiatry 2005;57: 1367–1373.
28 Polanczyk G, Faraone SV, Bau CH, Victor MM, Becker K, Pelz R, Buitelaar JK, Franke B, Kooij S, van der Meulen E, Cheon KA, Mick E, Purper-Ouakil D, Gorwood P, Stein MA, Cook EH Jr, Rohde LA: The impact of individual and methodological factors in the variability of response to methylphenidate in ADHD pharmacogenetic studies from four different continents. Am J Med Genet B Neuropsychiatr Genet 2008;147B:1419–1424.
29 Stein MA, McGough JJ: The pharmacogenomic era: promise for personalizing attention deficit hyperactivity disorder therapy. Child Adolesc Psychiatr Clin N Am 2008;17:475–490, xi–xii.
30 Jorgensen A, Alfirevic A: Pharmacogenetics and pharmacogenomics: adverse drug reactions. Pharmacogenomics 2008;9:1397–1401.
31 Wadelius M, Pirmohamed M: Pharmacogenetics of warfarin: current status and future challenges. Pharmacogenomics J 2007;7:99–111.
32 Lott DC, Kim SJ, Cook EH Jr, de Wit H: Dopamine transporter gene associated with diminished subjective response to amphetamine. Neuropsychopharmacology 2005;30:602–609.
33 American Psychological Association: PsycINFO homepage. 2008.
34 Volkow ND, Fowler JS, Wang G, Ding Y, Gatley SJ: Mechanism of action of methylphenidate: insights from PET imaging studies. J Atten Disord 2002; 6(suppl 1):S31–S43.
35 Melega WP, Williams AE, Schmitz DA, DiStefano EW, Cho AK: Pharmacokinetic and pharmacodynamic analysis of the actions of D-amphetamine and D-methamphetamine on the dopamine terminal. J Pharmacol Exp Ther 1995;274:90–96.
36 Cook EH Jr, Stein MA, Krasowski MD, Cox NJ, Olkon DM, Kieffer JE, Leventhal BL: Association of attention-deficit disorder and the dopamine transporter gene. Am J Hum Genet 1995;56:993–998.
37 VanNess SH, Owens MJ, Kilts CD: The variable number of tandem repeats element in DAT1 regulates in vitro dopamine transporter density. BMC Genet 2005;6:55.
38 Winsberg BG, Comings DE: Association of the dopamine transporter gene (DAT1) with poor methylphenidate response. J Am Acad Child Adolesc Psychiatry 1999;38:1474–1477.
39 Stein MA, Waldman ID, Sarampote CS, Seymour KE, Robb AS, Conlon C, Kim SJ, Cook EH: Dopamine transporter genotype and methylphenidate dose response in children with ADHD. Neuropsychopharmacology 2005;30:1374–1382.
40 Joober R, Grizenko N, Sengupta S, Amor LB, Schmitz N, Schwartz G, Karama S, Lageix P, Fathalli F, Torkaman-Zehi A, Ter Stepanian M: Dopamine transporter 3′-UTR VNTR genotype and ADHD: a pharmaco-behavioural genetic study with methylphenidate. Neuropsychopharmacology 2007;32:1370–1376.
41 McGough J, McCracken J, Swanson J, Riddle M, Kollins S, Greenhill L, Abikoff H, Davies M, Chuang S, Wigal T, Wigal S, Posner K, Skrobala A, Kastelic E, Ghuman J, Cunningham C, Shigawa S, Moyzis R, Vitiello B: Pharmacogenetics of methylphenidate response in preschoolers with ADHD. J Am Acad Child Adolesc Psychiatry 2006;45:1314–1322.
42 McGough JJ, Mccracken JT, Loo SK, Manganiello M, Leung MC, Tietjens JR, Trinh T, Baweja S, Suddath R, Smalley SL, Hellemann G, Sugar CA: A candidate gene analysis of methylphenidate response in attention-deficit/hyperactivity disorder. J Am Acad Child Adolesc Psychiatry 2009, E-pub ahead of print.
43 Gruber R, Joober R, Grizenko N, Leventhal BL, Cook EH Jr, Stein MA: Dopamine transporter genotype and stimulant side effect factors in youth diagnosed with attention-deficit/hyperactivity disorder. J Child Adolesc Psychopharmacol 2009;19: 233–239.
44 Leddy JJ, Waxmonsky JG, Salis RJ, Paluch RA, Gnagy EM, Mahaney P, Erbe R, Pelham WE, Epstein LH: Dopamine-related genotypes and the dose-response effect of methylphenidate on eating in attention-deficit/hyperactivity disorder youths. J Child Adolesc Psychopharmacol 2009;19:127–136.
45 Kirley A, Lowe N, Hawi Z, Mullins C, Daly G, Waldman I, McCarron M, O'Donnell D, Fitzgerald M, Gill M: Association of the 480 bp DAT1 allele with methylphenidate response in a sample of Irish children with ADHD. Am J Med Genet B Neuropsychiatr Genet 2003;121:50–54.
46 Cheon KA, Ryu YH, Kim JW, Cho DY: The homozygosity for 10-repeat allele at dopamine transporter gene and dopamine transporter density in Korean children with attention deficit hyperactivity disorder: relating to treatment response to methylphenidate. Eur Neuropsychopharmacol 2005;15: 95–101.
47 Roman T, Szobot C, Martins S, Biederman J, Rohde LA, Hutz MH: Dopamine transporter gene and response to methylphenidate in attention-deficit/hyperactivity disorder. Pharmacogenetics 2002;12: 497–499.

48 Purper-Ouakil D, Wohl M, Orejarena S, Cortese S, Boni C, Asch M, Mouren MC, Gorwood P: Pharmacogenetics of methylphenidate response in attention deficit/hyperactivity disorder: association with the dopamine transporter gene (SLC6A3). Am J Med Genet B Neuropsychiatr Genet 2008;147:1425–1430.
49 Kereszturi E, Tarnok Z, Bognar E, Lakatos K, Farkas L, Gadoros J, Sasvari-Szekely M, Nemoda Z: Catechol-O-methyltransferase val158met polymorphism is associated with methylphenidate response in ADHD children. Am J Med Genet B Neuropsychiatr Genet 2008;147:1431–1435.
50 Bellgrove MA, Barry E, Johnson KA, Cox M, Daibhis A, Daly M, Hawi Z, Lambert D, Fitzgerald M, McNicholas F, Robertson IH, Gill M, Kirley A: Spatial attentional bias as a marker of genetic risk, symptom severity, and stimulant response in ADHD. Neuropsychopharmacology 2008;33:2536–2545.
51 Tharoor H, Lobos EA, Todd RD, Reiersen AM: Association of dopamine, serotonin, and nicotinic gene polymorphisms with methylphenidate response in ADHD. Am J Med Genet B Neuropsychiatr Genet 2008;147B:527–530.
52 van der Meulen EM, Bakker SC, Pauls DL, Oteman N, Kruitwagen CL, Pearson PL, Sinke RJ, Buitelaar JK: High sibling correlation on methylphenidate response but no association with DAT1-10R homozygosity in Dutch sibpairs with ADHD. J Child Psychol Psychiatry 2005;46:1074–1080.
53 Zeni CP, Guimaraes AP, Polanczyk GV, Genro JP, Roman T, Hutz MH, Rohde LA: No significant association between response to methylphenidate and genes of the dopaminergic and serotonergic systems in a sample of Brazilian children with attention-deficit/hyperactivity disorder. Am J Med Genet B Neuropsychiatr Genet 2007;144:391–394.
54 Langley K, Turic D, Peirce TR, Mills S, Van Den Bree MB, Owen MJ, O'Donovan MC, Thapar A: No support for association between the dopamine transporter (DAT1) gene and ADHD. Am J Med Genet B Neuropsychiatr Genet 2005;139:7–10.
55 Mick E, Biederman J, Spencer T, Faraone SV, Sklar P: Absence of association with DAT1 polymorphism and response to methylphenidate in a sample of adults with ADHD. Am J Med Genet B Neuropsychiatr Genet 2006;141:890–894.
56 Kooij JS, Boonstra AM, Vermeulen SH, Heister AG, Burger H, Buitelaar JK, Franke B: Response to methylphenidate in adults with ADHD is associated with a polymorphism in SLC6A3 (DAT1). Am J Med Genet B Neuropsychiatr Genet 2008;147B:201–208.
57 Brookes K, Xu X, Chen W, Zhou K, Neale B, Lowe N, Anney R, Franke B, Gill M, Ebstein R, Buitelaar J, Sham P, Campbell D, Knight J, Andreou P, Altink M, Arnold R, Boer F, Buschgens C, Butler L, Christiansen H, Feldman L, Fleischman K, Fliers E, Howe-Forbes R, Goldfarb A, Heise A, Gabriels I, Korn-Lubetzki I, Johansson L, Marco R, Medad S, Minderaa R, Mulas F, Muller U, Mulligan A, Rabin K, Rommelse N, Sethna V, Sorohan J, Uebel H, Psychogiou L, Weeks A, Barrett R, Craig I, Banaschewski T, Sonuga-Barke E, Eisenberg J, Kuntsi J, Manor I, McGuffin P, Miranda A, Oades RD, Plomin R, Roeyers H, Rothenberger A, Sergeant J, Steinhausen HC, Taylor E, Thompson M, Faraone SV, Asherson P: The analysis of 51 genes in DSM-IV combined type attention deficit hyperactivity disorder: association signals in DRD4, DAT1 and 16 other genes. Mol Psychiatry 2006;11:934–953.
58 Asherson P, Brookes K, Franke B, Chen W, Gill M, Ebstein RP, Buitelaar J, Banaschewski T, Sonuga-Barke E, Eisenberg J, Manor I, Miranda A, Oades RD, Roeyers H, Rothenberger A, Sergeant J, Steinhausen HC, Faraone SV: Confirmation that a specific haplotype of the dopamine transporter gene is associated with combined-type ADHD. Am J Psychiatry 2007;164:674–677.
59 Brookes KJ, Mill J, Guindalini C, Curran S, Xu X, Knight J, Chen CK, Huang YS, Sethna V, Taylor E, Chen W, Breen G, Asherson P: A common haplotype of the dopamine transporter gene associated with attention-deficit/hyperactivity disorder and interacting with maternal use of alcohol during pregnancy. Arch Gen Psychiatry 2006;63:74–81.
60 Heckers S, Konradi C: Synaptic function and biochemical neuroanatomy; in Martin A, Scahill L, Charney DS, Leckman JF (eds): Pediatric Psychopharmacology: Principles and Practice. New York, Oxford University Press, 2003, pp 20–32.
61 Asghari V, Sanyal S, Buchwaldt S, Paterson A, Jovanovic V, Van Tol HH: Modulation of intracellular cyclic amp levels by different human dopamine D4 receptor variants. J Neurochem 1995;65:1157–1165.
62 Van Tol HH, Wu CM, Guan HC, Ohara K, Bunzow JR, Civelli O, Kennedy J, Seeman P, Niznik HB, Jovanovic V: Multiple dopamine D4 receptor variants in the human population. Nature 1992;358:149–152.
63 McCracken JT, Smalley SL, McGough JJ, Crawford L, Del'Homme M, Cantor RM, Liu A, Nelson SF: Evidence for linkage of a tandem duplication polymorphism upstream of the dopamine D4 receptor gene (DRD4) with attention deficit hyperactivity disorder (ADHD). Mol Psychiatry 2000;5:531–536.

64 Tahir E, Yazgan Y, Cirakoglu B, Ozbay F, Waldman I, Asherson PJ: Association and linkage of DRD4 and DRD5 with attention deficit hyperactivity disorder (ADHD) in a sample of Turkish children. Mol Psychiatry 2000;5:396–404.
65 Hamarman S, Fossella J, Ulger C, Brimacombe M, Dermody J: Dopamine receptor 4 (DRD4) 7-repeat allele predicts methylphenidate dose response in children with attention deficit hyperactivity disorder: a pharmacogenetic study. J Child Adolesc Psychopharmacol 2004;14:564–574.
66 Seeger G, Schloss P, Schmidt MH: Marker gene polymorphisms in hyperkinetic disorder – predictors of clinical response to treatment with methylphenidate? Neurosci Lett 2001;313:45–48.
67 Cheon KA, Kim BN, Cho SC: Association of 4-repeat allele of the dopamine D4 receptor gene exon III polymorphism and response to methylphenidate treatment in Korean ADHD children. Neuropsychopharmacology 2007;32:1377–1383.
68 Sallee FR, Newcorn J, Allen AJ, Stein MA, McDougle C, Sethuraman S: Pharmacogenetics of atomoxetine: Relevance of DRD4. 51st Ann Meet Am Acad Child Adolesc Psychiatry, Washington, 2004.
69 Nestler EJ, Hyman SE, Malenka RC: Molecular Neuropharmacology: A Foundation for Clinical Neuroscience. New York, McGraw-Hill, 2001.
70 Arnsten AF, Dudley AG: Methylphenidate improves prefrontal cortical cognitive function through alpha2 adrenoceptor and dopamine D1 receptor actions: relevance to therapeutic effects in attention deficit hyperactivity disorder. Behav Brain Funct 2005;1:2.
71 Andrews GD, Lavin A: Methylphenidate increases cortical excitability via activation of alpha-2 noradrenergic receptors. Neuropsychopharmacology 2006;31:594–601.
72 Lario S, Calls J, Cases A, Oriola J, Torras A, Rivera F: MspI identifies a biallelic polymorphism in the promoter region of the alpha 2A-adrenergic receptor gene. Clin Genet 1997;51:129–130.
73 Deupree JD, Smith SD, Kratochvil CJ, Bohac D, Ellis CR, Polaha J, Bylund DB: Possible involvement of alpha-2A adrenergic receptors in attention deficit hyperactivity disorder: radioligand binding and polymorphism studies. Am J Med Genet B Neuropsychiatr Genet 2006;141B:877–884.
74 Polanczyk G, Zeni C, Genro JP, Guimaraes AP, Roman T, Hutz MH, Rohde LA: Association of the adrenergic alpha2A receptor gene with methylphenidate improvement of inattentive symptoms in children and adolescents with attention-deficit/hyperactivity disorder. Arch Gen Psychiatry 2007;64: 218–224.
75 da Silva TL, Pianca TG, Roman T, Hutz MH, Faraone SV, Schmitz M, Rohde LA: Adrenergic alpha2A receptor gene and response to methylphenidate in attention-deficit/hyperactivity disorder-predominantly inattentive type. J Neural Transm 2008;115: 341–345.
76 Cheon KA, Cho DY, Koo MS, Song DH, Namkoong K: Association between homozygosity of a G allele of the alpha-2A-adrenergic receptor gene and methylphenidate response in Korean children and adolescents with attention-deficit/hyperactivity disorder. Biol Psychiatry 2009;65:564–570.
77 Sun Z, Murry DJ, Sanghani SP, Davis WI, Kedishvili NY, Zou Q, Hurley TD, Bosron WF: Methylphenidate is stereoselectively hydrolyzed by human carboxylesterase CES1A1. J Pharmacol Exp Ther 2004;310:469–476.
78 Zhu HJ, Patrick KS, Yuan HJ, Wang JS, Donovan JL, DeVane CL, Malcolm R, Johnson JA, Youngblood GL, Sweet DH, Langaee TY, Markowitz JS: Two CES1 gene mutations lead to dysfunctional carboxylesterase 1 activity in man: clinical significance and molecular basis. Am J Hum Genet 2008;82: 1241–1248.
79 Nemoda Z, Angyal N, Tarnok Z, Gadoros J, Sasvari-Szekely M: Carboxylesterase 1 gene polymorphism and methylphenidate response in ADHD. Neuropharmacology 2009;57:731–733.
80 Lachman HM, Papolos DF, Saito T, Yu YM, Szumlanski CL, Weinshilboum RM: Human catechol-O-methyltransferase pharmacogenetics: description of a functional polymorphism and its potential application to neuropsychiatric disorders. Pharmacogenetics 1996;6:243–250.
81 Cheuk DK, Wong V: Meta-analysis of association between a catechol-O-methyltransferase gene polymorphism and attention deficit hyperactivity disorder. Behav Genet 2006;36:651–659.
82 Cheon KA, Jun JY, Cho DY: Association of the catechol-O-methyltransferase polymorphism with methylphenidate response in a classroom setting in children with attention-deficit hyperactivity disorder. Int Clin Psychopharmacol 2008;23:291–298.
83 Mattay VS, Goldberg TE, Fera F, Hariri AR, Tessitore A, Egan MF, Kolachana B, Callicott JH, Weinberger DR: Catechol O-methyltransferase val158-met genotype and individual variation in the brain response to amphetamine. Proc Natl Acad Sci USA 2003;100:6186–6191.
84 Holmes A, Hollon TR, Gleason TC, Liu Z, Dreiling J, Sibley DR, Crawley JN: Behavioral characterization of dopamine D5 receptor null mutant mice. Behav Neurosci 2001;115:1129–1144.

85 Lowe N, Kirley A, Hawi Z, Sham P, Wickham H, Kratochvil CJ, Smith SD, Lee SY, Levy F, Kent L, Middle F, Rohde LA, Roman T, Tahir E, Yazgan Y, Asherson P, Mill J, Thapar A, Payton A, Todd RD, Stephens T, Ebstein RP, Manor I, Barr CL, Wigg KG, Sinke RJ, Buitelaar JK, Smalley SL, Nelson SF, Biederman J, Faraone SV, Gill M: Joint analysis of the DRD5 marker concludes association with attention-deficit/hyperactivity disorder confined to the predominantly inattentive and combined subtypes. Am J Hum Genet 2004;74:348–356.

86 Maher BS, Marazita ML, Ferrell RE, Vanyukov MM: Dopamine system genes and attention deficit hyperactivity disorder: a meta-analysis. Psychiatr Genet 2002;12:207–215.

87 Li D, Sham PC, Owen MJ, He L: Meta-analysis shows significant association between dopamine system genes and attention deficit hyperactivity disorder (ADHD). Hum Mol Genet 2006;15:2276–2284.

88 Xu F, Gainetdinov RR, Wetsel WC, Jones SR, Bohn LM, Miller GW, Wang YM, Caron MG: Mice lacking the norepinephrine transporter are supersensitive to psychostimulants. Nat Neurosci 2000;3:465–471.

89 Carboni E, Silvagni A, Vacca C, Di Chiara G: Cumulative effect of norepinephrine and dopamine carrier blockade on extracellular dopamine increase in the nucleus accumbens shell, bed nucleus of stria terminalis and prefrontal cortex. J Neurochem 2006;96:473–481.

90 Berridge CW, Devilbiss DM, Andrzejewski ME, Arnsten AF, Kelley AE, Schmeichel B, Hamilton C, Spencer RC: Methylphenidate preferentially increases catecholamine neurotransmission within the prefrontal cortex at low doses that enhance cognitive function. Biol Psychiatry 2006;60:1111–1120.

91 Michelson D, Faries D, Wernicke J, Kelsey D, Kendrick K, Sallee FR, Spencer T: Atomoxetine in the treatment of children and adolescents with attention-deficit/hyperactivity disorder: a randomized, placebo-controlled, dose-response study. Pediatrics 2001;108:E83.

92 Yang L, Wang YF, Li J, Faraone SV: Association of norepinephrine transporter gene with methylphenidate response. J Am Acad Child Adolesc Psychiatry 2004;43:1154–1158.

93 Dlugos A, Freitag C, Hohoff C, McDonald J, Cook EH, Deckert J, de Wit H: Norepinephrine transporter gene variation modulates acute response to D-amphetamine. Biol Psychiatry 2007;61:1296–1305.

94 Ramoz N, Boni C, Downing AM, Close SL, Peters SL, Prokop AM, Allen AJ, Hamon M, Purper-Ouakil D, Gorwood P: A haplotype of the norepinephrine transporter (NET) gene SLC6A2 is associated with clinical response to atomoxetine in attention-deficit hyperactivity disorder (ADHD). Neuropsychopharmacology 2009;34:2135–2142.

95 Seeger G, Schloss P, Schmidt MH: Functional polymorphism within the promotor of the serotonin transporter gene is associated with severe hyperkinetic disorders. Mol Psychiatry 2001;6:235–238.

96 Lesch KP, Bengel D, Heils A, Sabol SZ, Greenberg BD, Petri S, Benjamin J, Muller CR, Hamer DH, Murphy DL: Association of anxiety-related traits with a polymorphism in the serotonin transporter gene regulatory region. Science 1996;274:1527–1531.

97 Hranilovic D, Stefulj J, Schwab S, Borrmann-Hassenbach M, Albus M, Jernej B, Wildenauer D: Serotonin transporter promoter and intron 2 polymorphisms: relationship between allelic variants and gene expression. Biol Psychiatry 2004;55:1090–1094.

98 MacKenzie A, Quinn J: A serotonin transporter gene intron 2 polymorphic region, correlated with affective disorders, has allele-dependent differential enhancer-like properties in the mouse embryo. Proc Natl Acad Sci USA 1999;96:15251–15255.

99 Lott DC, Kim SJ, Cook EH, de Wit H: Serotonin transporter genotype and acute subjective response to amphetamine. Am J Addict 2006;15:327–335.

100 Schiavo G, Stenbeck G, Rothman JE, Sollner TH: Binding of the synaptic vesicle v-SNARE, synaptotagmin, to the plasma membrane t-SNARE, SNAP-25, can explain docked vesicles at neurotoxin-treated synapses. Proc Natl Acad Sci USA 1997;94:997–1001.

101 Barr CL, Feng Y, Wigg K, Bloom S, Roberts W, Malone M, Schachar R, Tannock R, Kennedy JL: Identification of DNA variants in the SNAP-25 gene and linkage study of these polymorphisms and attention-deficit hyperactivity disorder. Mol Psychiatry 2000;5:405–409.

102 Brophy K, Hawi Z, Kirley A, Fitzgerald M, Gill M: Synaptosomal-associated protein 25 (SNAP-25) and attention deficit hyperactivity disorder (ADHD): evidence of linkage and association in the Irish population. Mol Psychiatry 2002;7:913–917.

103 Kustanovich V, Merriman B, McGough J, McCracken JT, Smalley SL, Nelson SF: Biased paternal transmission of SNAP-25 risk alleles in attention-deficit hyperactivity disorder. Mol Psychiatry 2003;8:309–315.

104 Mill J, Curran S, Kent L, Gould A, Huckett L, Richards S, Taylor E, Asherson P: Association study of a SNAP-25 microsatellite and attention deficit hyperactivity disorder. Am J Med Genet 2002;114:269–271.

105 Wilson MC: Coloboma mouse mutant as an animal model of hyperkinesis and attention deficit hyperactivity disorder. Neurosci Biobehav Rev 2000;24:51–57.

106 Staddon S, Arranz MJ, Mancama D, Mata I, Kerwin RW: Clinical applications of pharmacogenetics in psychiatry. Psychopharmacology (Berl) 2002;162:18–23.

107 Dring LG, Smith RL, Williams RT: The metabolic fate of amphetamine in man and other species. Biochem J 1970;116:425–435.

108 Markowitz JS, Patrick KS: Pharmacokinetic and pharmacodynamic drug interactions in the treatment of attention-deficit hyperactivity disorder. Clin Pharmacokinet 2001;40:753–772.

109 Meyer UA, Zanger UM: Molecular mechanisms of genetic polymorphisms of drug metabolism. Annu Rev Pharmacol Toxicol 1997;37:269–296.

110 Wandel C, Witte JS, Hall JM, Stein CM, Wood AJ, Wilkinson GR: Cyp3a activity in African American and European American men: population differences and functional effect of the CYP3A4*1B5′-promoter region polymorphism. Clin Pharmacol Ther 2000;68:82–91.

111 Michelson D, Read HA, Ruff DD, Witcher J, Zhang S, McCracken J: CYP2D6 and clinical response to atomoxetine in children and adolescents with ADHD. J Am Acad Child Adolesc Psychiatry 2007;46:242–251.

112 Van der Meulen EM, Bakker SC, Pauls DL, Sinnke RJ, Buitelaar J: A genome-wide quantitative trait locus analysis on methylphenidate response rate in Dutch sibpairs with attention-deficit/hyperactivity disorder. 16th World Congr Int Assoc Child Adolesc Psychiatry. Allied Professions, Berlin, 2004.

113 Mick E, Neale B, Middleton FA, McGough JJ, Faraone SV: Genome-wide association study of response to methylphenidate in 187 children with attention-deficit/hyperactivity disorder. Am J Med Genet B Neuropsychiatr Genet 2008;147:1412–1418.

114 Pelham WE, Millich R: Individual differences in response to Ritalin in class work and social behavior; in Greenhill L, Osman B (eds): Ritalin: Theory and Patient Management. New York, Mary Ann Liebert, 1991.

115 Brody AL, Mandelkern MA, Olmstead RE, Scheibal D, Hahn E, Shiraga S, Zamora-Paja E, Farahi J, Saxena S, London ED, McCracken JT: Gene variants of brain dopamine pathways and smoking-induced dopamine release in the ventral caudate/nucleus accumbens. Arch Gen Psychiatry 2006;63:808–816.

116 Li S, Kim KY, Kim JH, Kim JH, Park MS, Bahk JY, Kim MO: Chronic nicotine and smoking treatment increases dopamine transporter mRNA expression in the rat midbrain. Neurosci Lett 2004;363:29–32.

117 Gerasimov MR, Franceschi M, Volkow ND, Rice O, Schiffer WK, Dewey SL: Synergistic interactions between nicotine and cocaine or methylphenidate depend on the dose of dopamine transporter inhibitor. Synapse 2000;38:432–437.

118 Weinshilboum RM, Wang L: Pharmacogenetics and pharmacogenomics: development, science, and translation. Annu Rev Genomics Hum Genet 2006;7:223–245.

119 Pliszka SR, Crismon ML, Hughes CW, Corners CK, Emslie GJ, Jensen PS, McCracken JT, Swanson JM, Lopez M: The Texas children's medication algorithm project: revision of the algorithm for pharmacotherapy of attention-deficit/hyperactivity disorder. J Am Acad Child Adolesc Psychiatry 2006;45:642–657.

120 McGough JJ, Biederman J, Wigal SB, Lopez FA, McCracken JT, Spencer T, Zhang Y, Tulloch SJ: Long-term tolerability and effectiveness of once-daily mixed amphetamine salts (Adderall XR) in children with ADHD. J Am Acad Child Adolesc Psychiatry 2005;44:530–538.

121 Wilens T, Pelham W, Stein M, Conners CK, Abikoff H, Atkins M, August G, Greenhill L, McBurnett K, Palumbo D, Swanson J, Wolraich M: ADHD treatment with once-daily OROS methylphenidate: interim 12-month results from a long-term open-label study. J Am Acad Child Adolesc Psychiatry 2003;42:424–433.

122 Charach A, Ickowicz A, Schachar R: Stimulant treatment over five years: adherence, effectiveness, and adverse effects. J Am Acad Child Adolesc Psychiatry 2004;43:559–567.

123 Wigal T, Greenhill L, Chuang S, McGough J, Vitiello B, Skrobala A, Swanson J, Wigal S, Abikoff H, Kollins S, McCracken J, Riddle M, Posner K, Ghuman J, Davies M, Thorp B, Stehli A: Safety and tolerability of methylphenidate in preschool children with ADHD. J Am Acad Child Adolesc Psychiatry 2006;45:1294–1303.

124 Siegel A, Sapru HN: Essential Neuroscience, ed 1 (Revised). Philadelphia, Lippincott Williams & Wilkins, 2006.

125 Moron JA, Brockington A, Wise RA, Rocha BA, Hope BT: Dopamine uptake through the norepinephrine transporter in brain regions with low levels of the dopamine transporter: evidence from knock-out mouse lines. J Neurosci 2002;22:389–395.
126 Bajjalieh SM, Scheller RH: The biochemistry of neurotransmitter secretion. J Biol Chem 1995;270:1971–1974.
127 Todd RD, O'Malley KL: The dopamine DRD4 receptor gene: are duplications distracting? Trends Pharmacol Sci 2001;22:55–56.
128 Thapar A, O'Donovan M, Owen MJ: The genetics of attention deficit hyperactivity disorder. Human Molecular Genetics 2005;14:R275–R282.
129 Arnsten AF, Dudley AG: Methylphenidate improves prefrontal cortical cognitive function through alpha2 adrenoceptor and dopamine D1 receptor actions: relevance to therapeutic effects in attention deficit hyperactivity disorder. Behav Brain Funct 2005;1:2.
130 Lahdesmaki J, Sallinen J, MacDonald E, Scheinin M: Alpha2a-adrenoceptors are important modulators of the effects of D-amphetamine on startle reactivity and brain monoamines. Neuropsychopharmacology 2004;29:1282–1293.
131 Ichikawa M, Okamura-Oho Y, Okunishi R, Kanamori M, Suzuki H, Ritani A, Nitta H, Eguchi N, Urade Y, Hayashizaki Y: Expression analysis of genes responsible for serotonin signaling in the brain. Neurobiol Dis 2005;19:378–385.

Mark A. Stein, PhD
Institute for Juvenile Research, University of Illinois at Chicago
1747 W. Roosevelt Road, Rm 155
Chicago, IL 60008 (USA)
Tel. +1 312 996 5797, Fax +1 847 412 0910, E-Mail mstein@uic.edu

Schwab M, Kaschka WP, Spina E (eds): Pharmacogenomics in Psychiatry.
Adv Biol Psychiatry. Basel, Karger, 2010, vol 25, pp 103–117

Pharmacogenomics of Eating Disorders

Palmiero Monteleone · Mario Maj

Department of Psychiatry, University of Naples SUN, Naples, Italy

Abstract

Anorexia nervosa (AN), bulimia nervosa (BN) and binge-eating disorder (BED) are psychiatric disorders characterized by abnormal eating behaviors that often result in dramatic physical consequences for the patients. The etiology of eating disorders (EDs) is currently unknown; however, a strong genetic contribution is likely involved. In the last 10 years, several polymorphic variants of genes coding substances involved in the modulation of eating behavior and the regulation of ED-related psychopathological dimensions have been assessed for association with AN, BN and BED. Results have been generally inconsistent because of methodological flaws and differences between studies. Pharmacogenomic investigations have suggested a possible role of some serotonin-linked gene polymorphisms in predicting the response to combined treatments with selective serotonin reuptake inhibitors plus psychotherapy in BN. Similarly, genetic variants of the melanocortin-4 receptor gene have been found to predict the outcome of gastric banding surgery in BED patients. Pharmacogenomics is at an early stage in EDs and should be pursued. To fulfill this aim, future clinical treatment studies in EDs could include a systematic recruitment of DNA samples in order to perform screening of genotypic polymorphisms or mutations that could identify genetic variants associated with therapeutic outcome and/or side effects. This will be useful in the prevention and treatment of EDs.

According to the Diagnostic and Statistical Manual for Mental Disorders IV (DSM-IV) [1], eating disorders (EDs) are divided into 3 major types: anorexia nervosa (AN), bulimia nervosa (BN) and eating disorders not otherwise specified (which mainly includes binge-eating disorder; BED). AN is characterized by restricted eating, obsessive fears of being fat and the voluntary pursuit of thinness with an inability to maintain a normal healthy body weight (BW). Despite increasing emaciation and a BW <85% of the ideal, individuals with AN are dissatisfied with the perceived size and shape of their body, and engage in unhealthy behaviors to perpetuate BW loss or prevent BW gain. DSM-IV differentiates between 2 subtypes: AN restricting subtype (ANR) and AN binge-eating/purging subtype. The restricting subtype is characterized by behaviors of extreme and prolonged fasting and restraint. The binge-purge subtype is also defined by prolonged fasting, although it is punctuated by episodes of overeating followed by behaviors to compensate for weight gain such as self-induced vomiting, the misuse of laxatives, diuretics or enemas, and excessive physical exercise.

BN is characterized by recurrent episodes of uncontrolled binge eating coupled with inappropriate compensatory behaviors, such as vomiting, laxative abuse, food restriction and/or excessive exercising. These behaviors are engaged in to prevent BW gain, because of the patient's pathological fear of becoming fat. Generally, because of the ingestion of some amount of food in the course of bingeing, BN patients have a normal BW. Similarly to AN, BN is classified into 2 subtypes: BN purging subtype, where patients induce vomiting and/or abuse laxatives and/or diuretics as compensatory behaviors, and BN non-purging subtype, where patients engage in excessive exercise and/or food restriction to prevent BW gain. BED is characterized by recurrent episodes of uncontrolled binge eating, as in BN, but without inappropriate compensatory behaviors. As a consequence of the ingestion of large amounts of food during the binge episodes, individuals with BED are generally obese.

Epidemiology and Comorbidity of EDs

In the National Comorbidity Survey (USA), the lifetime prevalence of DSM-IV AN was 0.9% among women and 0.3% among men; the lifetime prevalence of DSM-IV BN was 1.5% among women and 0.5% among men; the lifetime prevalence of DSM-IV BED was 3.5% among women and 2.0% among men [2]. In non-Western countries, the prevalence rates of AN and BN are lower than in Western countries (0.002–0.9 vs. 0.1–5.7% for AN; 0.46–3.2 vs. 0.3–7.3% for BN, in female subjects), although these are gradually increasing [3].

Patients with AN and BN have a high comorbidity with other psychiatric diagnoses especially affective and anxiety disorders [4]. Full BED is significantly associated with bipolar disorder, major depressive disorder, most anxiety disorders, substance use disorders, body dysmorphic disorder, kleptomania, irritable bowel syndrome and fibromyalgia [5]. Gadalla and Piran [6] found (in their meta-analysis of the literature between 1985 and 2006) significant co-occurrence rates of alcohol use disorders ranging between small and medium sizes for all patterns of EDs except AN in women. The effect size for any ED was 0.38, for AN 0.09, for BN 0.46, and for BED 0.39.

Life expectancy is reduced by 25 years in those females who have suffered from AN since the age of 15 years [7]. EDs were placed 4th in terms of burden of disease (years of life lost through death or disability) in women aged 15–24 years with both physical and psychological comorbidity [8]. With a standardized mortality rate of 23.14 [9], EDs have the highest mortality rate of any psychiatric disorder. Mortality can be due to suicide, medical complications of malnutrition or complicating comorbid medical disorders.

Genetics of EDs

The etiopathogenesis of EDs is thought to be multifactorial with psychological, social and biological factors allegedly being involved, although the exact role played by each of these is not yet completely known [10]. At present, it is widely accepted that genetic factors are responsible for the transmission of a biological vulnerability to the disorders as clearly demonstrated by epidemiology, family and twin studies.

A significantly higher frequency of AN or BN has been reported in relatives of probands with an ED as compared with relatives of healthy controls, which suggests a familial aggregation for AN and BN [11]. Additional studies have demonstrated co-aggregation of AN and BN, with relative risks for AN of 11.3 and 12.3, respectively, and relative risks for BN of 4.2 and 4.4, respectively, in first-degree female relatives of probands with AN or BN [12]. These data suggest that familial vulnerabilities could be partly shared in different EDs and that specific genes can predispose individuals to both EDs.

Most twin studies have shown a higher concordance rate of AN or BN in monozygotic twins than in dizygotic twins, which implies that genetic factors (more than shared familial environment) may explain why EDs run in families. The heritability estimates from these studies have been calculated to range from 48 to 88% in AN and from 28 to 83% in BN [11, 13]. Klump et al. [14] explored the influence of genetic and environmental factors on ED-related behaviors in pre-pubertal and post-pubertal 11-year-old twin girls as compared to a 17-year-old post-pubertal twin cohort and found that in pre-pubertal twins no significant influence of additive genetic factors was evident, yet common environmental factors were important. Contrary to this, in both post-pubertal 11-year-old and 17-year-old girls, genetic effects were significant, whereas shared environment was not. These findings suggest that genes of vulnerability to EDs are likely activated during puberty.

Molecular genetic studies aiming to identify chromosomal regions and genes of vulnerability in EDs use 2 methodologies: linkage and association (i.e. case-control or within-family) designs. The first wide-genome scan performed in a heterogeneous sample of individuals with broadly defined EDs did not detect susceptibility loci; however, when only pairs exhibiting the classic ANR phenotype were analyzed, a significant linkage was found between the chromosomal 1p33–p36 region and ANR [15]. A subsequent more detailed linkage analysis confirmed the existence of a susceptibility locus for ANR on chromosome 1p33–p36 and found that 2 candidate genes in this region, *HTR1D* and *OPRD1* encoding serotonin (5-HT)-1D and opioid delta-1 receptors, were significantly associated to ANR [16]. A genome-wide linkage analysis based on a large cohort of families in which at least 2 biological relatives were affected by BN reported a significant linkage with the chromosome 10p13–p12 region and a suggestive linkage with chromosome 14q [17]. To our knowledge, no linkage study has been performed for BED.

More genetic association studies than linkage studies have been performed on EDs. This has been made possible by the large amount of data coming from our understanding of the biological mechanisms controlling the physiology of food intake, appetite, satiety and BW. In the last 20 years, central and/or peripheral neurotransmitters, hormones and peptides have been identified that regulate eating behavior and, in some cases, psychopathological dimensions associated to EDs [18]. Therefore, the genes involved in the biosynthesis and/or degradation of those substances and their receptors have been selected as candidate genes, and mutations, variations and polymorphisms of those genes appear to be of particular interest, especially if they affect either the protein structure/function or expression. A critical appraisal of association studies of candidate gene polymorphisms in AN, BN and BED has been recently published by the authors [19] and an overview of those studies is provided in tables 1–3.

Treatment of EDs

The treatment of EDs is based on a multidisciplinary approach that combines nutritional rehabilitation, psychotherapies and drug treatments. Psychotherapies usually involve several strategies, including cognitive-behavioral and family therapies. Not least because of the broad spectrum of psychiatric disorders which have substantial comorbidity with EDs and their possible effect on eating behavior, a lot of psychopharmacological agents including antidepressants, antipsychotics, antiepileptics, antihistaminics and other pharmacological compounds have been investigated in the treatment of EDs.

In the NICE guidelines [20], pharmacotherapy is not seen as first choice for EDs, but is mentioned as an adjunct to psychological therapies or to treat physical or comorbid psychological problems. For AN, the NICE guidelines mention medication as disappointing in influencing the core symptoms of the disorder, promoting weight gain or reducing associated mood disturbance. In general, selective serotonin reuptake inhibitors (SSRIs) have been proved to be ineffective in patients with AN who are underweight, whereas they have been demonstrated to be helpful in relapse prevention in weight-restored patients [21], although this has been recently questioned [22]. For BN and BED, the NICE guidelines see some evidence that antidepressants, particularly SSRIs, contribute to the cessation of binge eating and purging.

Pharmacogenomics of EDs

Clinical trials and the clinical practice show that an appreciable proportion of ED patients do not respond adequately to treatments, while others may have adverse effects that can be responsible for the early discontinuation of therapy. It is reasonable

Table 1. Overview of genetic association studies on AN

Candidate gene	Analyzed polymorphism	Number of studies	Results
Serotonin transporter	44-bp Del/Ins (promoter)	7	nominal association of the S allele detected in 2 studies, but not confirmed in 5 others; a meta-analysis concluded that the S allele may represent a moderate but significant risk for AN
5-HT$_{2A}$ receptor	−1438G/A (promoter)	15	significantly higher frequency of the AA genotype and the A allele of the −1438G/A polymorphism was found by some studies, but not confirmed by others; in 4 of the positive studies, the A allele was associated specifically to ANR
	Thr25Asn	2	no significant association
	His452Tyr	2	
	102T/C	1	
	516T/C	1	
5-HT$_{2C}$ receptor	Cys23Ser	3	2 out of 3 studies reported higher frequencies of the Ser23Ser genotype and the Ser23 allele; Ser23 allele was suggested as predisposing young women to lose weight through reducing food intake
5-HT$_{1D\beta}$ receptor	Phe124Cys	1	some SNPs and the haplotypes −1123T>C/1080C>T and 1080C>T/2190A>G of the *5-HT$_{1D\beta}$* receptor gene were significantly associated with AN or ANR
	−1123T>C	1	
	−628T>C	1	
	1080C>T	1	
	2190A>G	1	
	rs652783	1	
	rs604030	1	
	rs674386	1	
	rs856510	1	
5-HT$_7$ receptor	Pro279Leu	1	no significant association
Tryptophan hydroxylase-1	T1095C	1	no significant association
Dopamine transporter (DAT1)	VNTR	1	higher frequency of short alleles (7 and 9 repeats) as compared to long alleles (10 and 11 repeats) found in ANBP
Dopamine D2 receptor (DRD2)	−141 Indel-/C	1	significant associations were found between the 725bp3'G>T and 10620C>T SNPs and ANBP; moreover, the haplotype Indel+939C>T was significantly associated to both AN and ANR; the haplotypes Indel+957C>T and 939C>T+725bp3'G>T were significantly associated to AN, and the haplotype 939C>T+10520C>T was significantly associated to ANR
	2730T>C	1	
	932C>G	1	
	939C>T	1	
	957C>T	1	
	725bp3'G>T	1	
	10620C>T	1	
	TaqA1	1	no significant association
Dopamine D3 receptor (DRD3)	Bal-I	1	no significant association
Dopamine D4 receptor (DRD4)	3-bp deletion	1	C allele of the D4pr C(521)T SNP was preferentially transmitted to AN individuals; the D4pr C(521)T SNP, the D4pr 120 repeat and several 2-, 3-, 4- and 5-locus haplotypes were significantly associated with AN (with some differences between ANR and ANBP)
	48-bp repeat	1	
	D4pr C(521)T	1	
	D4pr C(616)G	1	
	D4pr A(809)G	1	
	D4pr 120 repeat	1	
	D4 exon III repeat	1	

Table 1 (continued)

Candidate gene	Analyzed polymorphism	Number of studies	Results
Noradrenaline transporter (NET)	4-bp Del/Ins (promoter)	2	significant association found in ANR but not in ANBP patients
β3-adrenergic receptor	Trp64Arg	1	no significant association
Catechol-O-methyltransferase (COMT)	Val158Met (472G/A)	4	nominal association of the ValVal genotype and the Val allele with AN has been shown in 2 studies, but not confirmed in a large case-control study; similarly, a preferential transmission of the Val allele was detected in a sample of 66 ANR trios, but not confirmed in a larger study including 372 AN trios
	–1219A/G	1	no significant association
	186C/T	1	
	408C/G	1	significant association found in the ANR but not in the ANBP
ARVCF	826InsC	1	no significant association
	659C/T	1	
	524T/C	1	significant association found in ANR but not in ANBP
Monoamine oxidase A (MAOA)	MAOA-uVNTR	1	no significant association
Brain-derived neurotrophic factor (BDNF)	196G/A (Val66Met)	8	the Val66Met SNP of the *BDNF* gene was found quite consistently although not specifically linked to ANR
	–270C/T	5	no significant association found in 5 studies, but AN patients with the T allele had higher levels of persistence and harm avoidance
Neurotrophic tyrosin kinase receptor 2 (NTRK2)	–69C>G	1	nominal association found in ANBP but not in ANR
	IVS2+40C>T	1	no significant association found for any of these SNPs, but the C-A-insC haplotype was strongly associated with ANBP
	IVS13+40G>A	1	
	IVS17+125T>C	1	
	IVS18+13G>A	1	
	2785–2785insC	1	
Neuropeptide Y Y_1 receptor	*Pst*I	1	no significant association
Neuropeptide Y Y_5 receptor	Gly426Gly	1	no significant association
Agouti-related protein (AGRP)	G760A	2	G760A and G526A, but not C659T, were found nominally associated to AN and the mutant allele was preferentially transmitted to ANBP offspring
	G526A	1	
	C659T	1	
Opioid receptor delta-1	80T>G	1	47821A>G SNP and the haplotypes 8214T>C/47821A>G and 80A>G/8214T>C/23340A>G/47821T>G/51502A>T were found nominally associated with AN
	8214T>C	1	
	23340A>G	1	
	47821A>G	1	
	51502A>T	1	
	rs569356	1	statistically significant association of these 3 SNPs was found in ANR but not in ANBP
	rs521809	1	
	rs4654327	1	
	rs204055	1	
	rs204047	1	no significant association
	rs2298896	1	

Table 1 (continued)

Candidate gene	Analyzed polymorphism	Number of studies	Results
Cannabinoid receptor 1 (CNR1)	AAT 7,9-15 repeats	2	13-repeat allele was found preferentially transmitted in ANBP patients while the 14-repeat allele was found preferentially transmitted in ANR patients in 1 study, but not confirmed in another
	–22,959A/G	1	no significant association or transmission was found in 91 German AN trios
	–6,274A/T	1	
	–6,215T/G	1	
	–5,489T/C	1	
	–1,359G/A	1	
Fatty acid amide hydrolase (FAAH)	–272G/A	1	no significant association or transmission found in 91 German AN trios
	10,741C/A	1	
	11,966G/A	1	
	13,883G/A	1	
	19,542C/A	1	
N-acylethanolamine-hydrolyzing acid amidase (NAAA)	368A/G	1	no significant association or transmission found in 91 German AN trios
	9,263A/T	1	
	19,229G/T	1	
Ghrelin	Arg51Gln	3	no significant association
	Gln90Leu	3	
	Leu72Met	4	family trios study reported a significant association to ANBP and a preferential transmission of the Met allele and of the haplotype 90Gln/72Met to ANBP offspring; 2 case-control studies and a large family trios study did not confirm these results
	3056T>C	1	no significant association
	3083A>G	1	
	3615A>C	1	
Growth hormone secretagogue receptor (ghrelin receptor)	171T/C	1	no significant association
Cholecystokinin (CCK)	rs11129946	1	nominal association
	rs6791019	1	no significant association
	rs7611677	1	
	rs6809785	1	
	rs6801844	1	
CCK-A receptor	–81A>G	1	no significant association
	–128G>T	1	
Leptin	–1387 G/A	1	no significant association
Leptin receptor	Gln223Arg	1	no significant association
	Lys109Arg	1	
	Lys656Asn	1	
Adiponectin	45T>G	1	no significant association
	276G>T	1	
Resistin	62G>A	1	no significant association
	180C>G	1	

Table 1 (continued)

Candidate gene	Analyzed polymorphism	Number of studies	Results
Uncoupling protein 2, 3 (UCP2/UCP-3)	D11S911	2	allele 13 of D11S911 microsatellite marker was found significantly over represented in AN in 1 study, but not confirmed in an another
	D11S916 –866G/A (UCP-2) –55C/T (UCP-3)	2 1 1	no significant association
Tumor necrosis factor-α (TNFα)	–1031T>C –863C>A –857C>T 308G/A	1 1 1 1	no significant association
Phospholipase A2	*intPLA2*	1	no significant association
Estrogen receptor-1 (ESR1)	ESR1-PvuI ESR1-XbaI dinucleotide repeat	1 1 1	no significant association
Estrogen receptor-2 (ESR2 or ESR-β)	1082G>A	2	significant association with AN detected in 2 independent case-control studies
	1730A>G dinucleotide repeat	2 1	no significant association
Calcium-activated potassium channel (KCNN3)	CAG repeat	2	alleles longer than 19 repeats were found more frequently in AN and preferentially transmitted to offspring
Circadian locomotor output cycles kaput (CLOCK)	3111T/C	1	AN subjects with at least 1 copy of the C allele exhibited a minimum past BW significantly lower than those with T/T genotype

For references, see Monteleone and Maj [19]. ANBP = AN binge-purging subtype; ARVCF = armadillo repeat gene deleted in velocardiofacial syndrome; MAOA-uVNTR = MAOA-upstream variable number of tandem repeats.

to assume that the availability of predictors of treatment outcomes would be helpful to clinicians in order to select the most adequate treatment for a given patient. In this regard, pharmacogenomics may be a promising strategy to select the best pharmacological treatment.

The serotonin transporter protein (5-HTT) represents the prime target of SSRIs. The human 5-HTT is coded by a gene located on chromosome 17q11.1–17q.12. A polymorphism in the promoter region of the *5-HTT* gene (*5-HTTLPR*) consisting of a 44-bp deletion (short or S variant) or insertion (long or L variant) has been shown to be endowed with functional consequences as the S form is associated with a lower transcriptional activity and a reduced 5-HT reuptake efficiency than the L isoform [23]. It has been quite consistently shown that the S allele of the 5-HTTLPR is associated with poorer response to SSRIs in patients with major depression [24]. Given this background, the role of the 5-HTTLPR polymorphism in predicting the response to SSRIs

Table 2. Overview of genetic association studies in BN

Candidate gene	Analyzed polymorphism	Number of studies	Results
5-HT transporter	44-bp Del/Ins (promoter)	4	1 study showed a positive association between the S allele of the *5-HTTLPR* polymorphism and BN; another found a higher frequency of the L allele in BN females; 2 studies did not report any significant association
5-HT$_{2A}$ receptor	–1438G/A (promoter)	8	3 studies reported a significant association of the polymorphic A allele with BN, whereas 5 others did not; 1 study found a significant association of BN with the G allele instead of the A allele
	Thr25Asn His452Tyr 102T/C 516T/C	1 1 1 1	no significant association
5-HT$_{2C}$ receptor	Cys23Ser	2	no significant association
5-HT$_{1D\beta}$ receptor	G861C	2	no significant association was detected, but bulimic individuals with GG genotype had a significantly lower minimum lifetime BMI and severer comorbid obsessive-compulsive disorder compared to those with CC genotype
Tryptophan hydroxylase-1	A218C	1	the A allele was associated with a severer bulimic symptomatology and higher levels of harm avoidance
Dopamine transporter (DAT1)	VNTR	1	no significant association
Dopamine D2 receptor (DRD2)	TaqA1	1	no significant association
β3-adrenergic receptor	Trp64Arg	1	no significant association
Catechol-O-methyl-transferase (COMT)	Val158Met (472G/A)	1	no significant association was found in a study including only 28 BN individuals
Brain-derived neurotrophic factor (BDNF)	196G/A (Val66Met)	5	higher frequency of the AA/AG genotype and the A-allele was found in 389 BN individuals recruited from 3 European countries, but this was not confirmed in a subsequent family trios study and in 2 other small studies, although in 1 of them a significant association was found with BNNP type
	–270C/T	3	no significant association detected in 3 studies; however, BN individuals carrying the T allele exhibited an earlier age at onset of weight loss and higher maximum BMI
Neurotrophic tyrosin kinase receptor 2 (NTRK2)	–69C>G IVS2+40C>T IVS13+40G>A IVS17+125T>C IVS18+13G>A 2785–2785insC	1 1 1 1 1 1	IVS13+40G>A and IVS17+125T>C SNPs were found significantly associated with BN
Ghrelin	Arg51Gln Gln90Leu	3 2	no significant association was found in 2 studies, but the haplotype Gln90/Leu72/Arg51 was preferentially transmitted to BN offspring
	Leu72Met	3	significant association with BNP in 1 study
	3056T>C	1	significant association with BNP
	3083A>G 3615A>C	1 1	no significant association
Growth hormone secretagogue receptor (ghrelin receptor)	171T/C	1	significant association

Table 2 (continued)

Candidate gene	Analyzed polymorphism	Number of studies	Results
CCK-A receptor	–81A>G –128G>T	1 1	no significant association
Leptin	–1387 G/A	1	no significant association
Estrogen receptor-2 (ESR2 or ESR-b)	1082G>A	2	no significant association
	1730A>G	2	significant association detected in 1 case-control study, but not in another
	ERβ cx+56G>A	1	no significant association
Circadian locomotor output cycles kaput (CLOCK)	3111T/C	1	BN subjects with at least 1 copy of the C allele exhibited a minimum past BW significantly lower than those with T/T genotype

For references, see Monteleone and Maj [19]. BNP = Bulimia nervosa purging subtype; BNNP = bulimia nervosa non-purging subtype; CCK = cholecystokinin; VNTR = variable number of tandem repeats.

has also been explored in patients with BN. One study found that the S form of the 5-HTTLPR was associated with a poorer outcome of SSRI therapy in BN women undergoing a 12-week treatment with different SSRIs plus nutritional counseling in a naturalistic setting [25]. Indeed, as compared to patients with LL genotype, bulimic subjects carrying at least 1 copy of the S allele had a 23.33-fold reduced probability to get response, defined as a >50% decrease in the weekly frequency of binge-purging episodes. BN patients with LL genotype had a 10.66-fold increase in the probability of remission, defined as a complete absence of binge-purging episodes. Another study, however, did not confirm such an association between the 5-HTTLPR polymorphism and treatment outcome in BN individuals undergoing a 6-week treatment with 1 of 4 different SSRIs or placebo plus a standardized intensive cognitive-behavioral treatment program [26]. However, in this study, the efficacy of drug treatment was evaluated as the percent improvement in the Yale-Brown-Cornell Eating Disorders Scale, which is an instrument designed to assess food-related obsessions and compulsions and does not provide any specific measure of bulimic symptomatology.

Recently, Steiger et al. [27] explored the role of both the 5-HTTLPR and the –1438G/A SNP of the *5-HT$_{2A}$* receptor gene in the treatment outcome of women with bulimia-spectrum EDs undergoing a multimodal intervention strategy, which involved individual psychotherapy (mainly cognitive-behavioral therapy) and pharmacotherapy (mainly SSRIs or the serotonin-norepinephrine reuptake inhibitor venlafaxine alone or in combination with mood stabilizers, anxiolytics or other antidepressants) for 8 months in a naturalistic setting. Similarly to the 5-HTTLPR, the –1438G/A polymorphism in the promoter region of the *5-HT$_{2A}$* receptor gene is en-

Table 3. Overview of genetic association studies in BED

Candidate gene	Analyzed polymorphism	Number of studies	Results
5-HT transporter	44-bp Del/Ins (promoter)	1	higher frequency of LL genotype and L allele was detected in a relatively small group of obese women with BED as compared to normal weight healthy females
5-HT$_{2A}$ receptor	–1438G/A (promoter)	1	no significant association
5-HT$_{2C}$ receptor	Cys23Ser	1	no significant association
Dopamine transporter (DAT1)	VNTR	1	no significant association found; however, methylphenidate induced greater appetite suppression in BED subjects with at least one 9-repeat allele than in those with the 10/10 repeat genotype and in controls with at least one 9-repeat allele
Brain-derived neurotrophic factor (BDNF)	196G/A (Val66Met)	1	no significant association was found; however, subjects carrying the A/A genotype exhibited BITE severity scores and a weekly frequency of bingeing significantly higher than A/G and G/G genotypes
Melanocortin-4 receptor (MC4R)	different mutations	7	in a sample of severely obese patients, all those carrying *MC4R* gene variants fulfilled the DSM-IV criteria for BED, suggesting that *MC4R* variants constitute a genetic vulnerability for BED; in subsequent studies, no increased rates of BED were reported in other groups of *MC4R* mutation carriers extrapolated from samples of obese patients nor higher frequency of *MC4R* mutations were detected in primarily diagnosed BED patients with or without obesity
Fatty acid amide hydrolase (FAAH)	385C>A	1	higher frequencies of the CA genotype and the A allele were detected in a sample of obese individuals with or without BED and resulted correlated to the presence of overweight/obesity but not to the occurrence of BED
Ghrelin	Leu72Met Arg51Gln	1 1	significant association was found no significant association
Circadian locomotor output cycles kaput (CLOCK)	3111T/C	1	no significant association was found; however, overweight/obese subjects carrying the TT genotype had higher BMI scores compared to those carrying the CC genotype

For references, see Monteleone and Maj [19]. VNTR = Variable number of tandem repeats; BITE = Bulimia Investigation Test Edinburgh.

dowed with functional consequences as the functional activity of the promoter and the 5-HT$_{2A}$ receptor activation have been reported to be lower for the G allele and higher for the A allele [28]. In line with the results of Monteleone et al. [25], Steiger et al. [27] found that women with bulimia-spectrum EDs carrying the 5-HTTLPR low functional allele or the low function allele of the –1438G/A SNP of the *5-HT$_{2A}$* receptor gene showed less reduction in weekly frequency of binge eating after 8 months of treatment (in those patients carrying both SNPs), slower decrease in anxiety and depression symptomatology at 4 months (in 5-HTTLPR S-allele carriers) and absence of improvements on measures of impulsivity at 8 months (in –1438G/A G-allele carriers).

The pharmacogenomic approach has been pursued also in patients with BED. In this regard, polymorphic variants of the dopamine transporter *(DAT1)* gene and mutations of the melanocortin-4 receptor *(MC4R)* gene have been studied as putative predictors of outcome to both pharmacological and surgical treatments in BED patients with or without obesity. The DAT protein is a critical regulator of synaptic dopamine. The *DAT1* gene, encoding the DAT protein, is polymorphic and in most human beings it occurs with greatest frequency in the 9- and 10-repeat forms [29]. The 10-repeat variable number of tandem repeats polymorphism has been shown to be associated with an approximate 50% increase in DAT binding sites as compared to the 9-repeat allele [30]. One study showed that BED patients carrying the 9-repeat allele of the *DAT1* gene had a greater appetite suppression following methylphenidate administration compared to controls with the same allele or to both patients and controls with 10/10 repeat *DAT1* genotype, which suggests a better putative therapeutic effect of the appetite suppressant drugs in the *DAT1* 9-repeat allele carrier BED patients [31].

Melanocortin-4 receptors (MC4R) in the hypothalamus are the target of anorexigenic melanocortins derived from proopiomelanocortin. Several loss-of-function mutations in the *MC4R* gene have been described, and Branson et al. [32] found that in a sample of severely obese study participants, all those carrying *MC4R* gene variants fulfilled the DSM-IV criteria for BED, suggesting that *MC4R* variants constitute a genetic vulnerability for BED. This hypothesis was questioned because most of the *MC4R* mutations reported by Branson et al. [32] are functionally inactive, and no increased rates of BED have been reported in other groups of *MC4R* mutation carriers extrapolated from samples of obese patients, nor has a higher frequency of *MC4R* mutations been detected in primarily diagnosed BED patients with or without obesity [19]. Nevertheless, in one study, obese patients with or without BED who underwent laparoscopic gastric banding treatment and were screened for *MC4R* variants were found to have different outcomes after the gastric banding treatment according to their *MC4R* genotypes and the presence or absence of BED [33]. Indeed, all *MC4R* variant carriers had BED and during a 3-year follow-up lost less weight, showed less improvement in metabolic syndrome, had dilated esophagi, more vomiting and 5 times more gastric complications than non-carriers. Overall outcome was poorest in *MC4R* variant carriers with BED, better in non-carriers with BED and best in non-carriers without BED.

To sum up, so far only 3 studies have been performed to assess the pharmacogenomics of BN and all of them have some methodological limitations (naturalistic design, relatively low number of participants, use of rating instruments not specific for BN psychopathology). Regardless, there remains an implication that the 5-HTTLPR and the –1438G/A low-function alleles predict lesser or slower response to combined pharmacological/psychotherapeutic treatments in BN patients. The *DAT1* gene polymorphisms and mutations of the *MC4R* gene seem to be promising candidates to predict treatment outcome in BED patients.

Conclusions and Future Perspectives

Epidemiology and genetic studies converge on the fact that EDs display a strong genetic component. In the last 10 years, variations in numerous candidate genes have been assessed for an association with AN, BN and BED. Results have been often inconsistent, since the majority of association studies has been performed on small subject samples, different ethnic populations, differently diagnosed ED patients, and suffered from insufficient statistical power, lack of correction for multiple testing, genetic heterogeneity and stratification. Rather, in these disorders, SNPs have been found frequently associated with ED-related phenotypic traits and not to AN, BN or BED as currently categorized in DSM-IV. Although these results have not been confirmed in large study groups, they underline the importance, in future studies, to focus on more homogeneous subgroups, either relying on specific ED traits or identifying endophenotypes.

Pharmacogenomic studies are at an early stage in EDs and should be pursued. To fulfill this aim, future clinical treatment studies in EDs could include a systematic recruitment of DNA samples in order to perform the screening of genotypic polymorphisms or mutations that could identify genetic variants associated with therapeutic and/or side effects. These studies should explore not only genes of the neurotransmitter systems and of the various peptides specifically involved in the modulation of eating behavior, energy homeostasis and the core psychopathology of ED patients, but also variants of those liver enzymes (CYP450 system), which metabolize the drugs used in the treatment of EDs. It is known that hepatic CYP enzymes are encoded by polymorphic genes, which generate proteins with different enzymatic efficiency. Therefore, the screening of CYP gene variants may enable the clinician to prevent administration of high CYP-metabolized drug doses in subjects who express CYP variants with reduced activity (poor metabolizers) in order to minimize the occurrence of possible adverse and toxic effects. Moreover, potential inefficacy will be also predicted in rapid metabolizers. Such kind of studies are at present completely lacking in EDs.

In conclusion, genetic studies of EDs clearly are in an early phase. Increasing knowledge of the mechanisms regulating BW and eating behavior as well as a more homogeneous characterization of clinical phenotypes will help to identify the genes likely involved in the heritable transmission of the biological vulnerability to these serious and debilitating conditions. Results in this field combined with the identification of genetic predictors of treatment outcomes will help clinicians to plan more effective preventive strategies and treatment programs.

References

1 American Psychiatric Association: Diagnostic and Statistical Manual of Mental Disorders, ed. 4. Washington, American Psychiatric Association, 1994.

2 Hudson JI, Hiripi E, Pope HG Jr, Kessler RC: The prevalence and correlates of eating disorders in the National Comorbidity Survey Replication. Biol Psychiatry 2007;61:348–358.

3 Makino M, Tsuboi K, Dennerstein L: Prevalence of eating disorders: a comparison of Western and non-Western countries. Med Gen Med 2004;6:49.

4 Halmi KA, Eckert E, Marchi P, Sampugnaro V, Apple R, Cohen J: Comorbidity of psychiatric diagnoses in anorexia nervosa. Arch Gen Psychiatry 1991; 48:712–718.

5 Javaras KN, Pope HG, Lalonde JK, Roberts JL, Nillni YI, Laird NM, Bulik CM, Crow SJ, McElroy SL, Walsh BT, Tsuang MT, Rosenthal NR, Hudson JI: Co-occurrence of binge eating disorder with psychiatric and medical disorders. J Clin Psychiatry 2008;69:266–273.

6 Gadalla T, Piran N: Co-occurrence of eating disorders and alcohol use disorders in women: a meta analysis. Arch Womens Ment Health 2007;10:133–140.

7 Harbottle EJ, Birmingham CL, Sayani F: Anorexia nervosa: a survival analysis. Eat Weight Disord 2008;13:32–34.

8 Mathers CD, Vos ET, Stevenson CE, Begg SJ: The Australian Burden of Disease Study: measuring the loss of health from diseases, injuries and risk factors. Med J Aust 2000;172:592–596.

9 Harris EC, Barraclough B: Suicide as an outcome for mental disorders: a meta-analysis. Br J Psychiatry 1997;170:205–228.

10 Gillberg C, Rastam M: Etiology in anorexia nervosa; in Hoek HW, Treasure JL, Katzman MA (eds): Neurobiology in the Treatment of Eating Disorders. Chichester, Wiley & Sons, 1998, pp 127–141.

11 Gorwood P, Kipman A, Foulon C: The human genetics of anorexia nervosa. Eur J Pharmacol 2003; 480:163–170.

12 Strober M, Freeman R, Lampert C, Diamond J, Kaye W: Controlled family study of anorexia nervosa and bulimia nervosa: evidence of shared liability and transmission of partial syndromes. Am J Psychiatry 2000;157:393–401.

13 Bulik CM, Sullivan PF, Wade TD, Kendler KS: Twin studies of eating disorders: a review. Int J Eat Disord 2000;27:1–20.

14 Klump KL, McGue M, Iacono WG: Differential heritability of eating attitudes and behaviors in prepubertal versus pubertal twins. Int J Eat Disord 2003;33:287–292.

15 Grice DE, Halmi KA, Fichter MM, Strober M, Woodside DB, Treasure JT, Kaplan AS, Magistretti PJ, Goldman D, Bulik CM, Kaye WH, Berrettini WH: Evidence for a susceptibility gene for anorexia nervosa on chromosome 1. Am J Hum Genet 2002; 70:787–792.

16 Bergen AW, van den Bree MB, Yeager M, Welch R, Ganjei JK, Haque K, Bacanu S, Berrettini WH, Grice DE, Goldman D, Bulik CM, Klump K, Fichter M, Halmi K, Kaplan A, Strober M, Treasure J, Woodside B, Kaye WH: Candidate genes for anorexia nervosa in the 1p33–36 linkage region: serotonin 1D and delta opioid receptor loci exhibit significant association to anorexia nervosa. Mol Psychiatry 2003;8:397–406.

17 Bulik CM, Devlin B, Bacanu SA, Thornton L, Klump KL, Fichter MM, Halmi KA, Kaplan AS, Strober M, Woodside DB, Bergen AW, Ganjei JK, Crow S, Mitchell J, Rotondo A, Mauri M, Cassano G, Keel P, Berrettini WH, Kaye WH: Significant linkage on chromosome 10p in families with bulimia nervosa. Am J Hum Genet 2003;72:200–207.

18 Monteleone P, Castaldo E, Maj M: Neuroendocrine dysregulation of food intake in eating disorders. Regul Pept 2008;149:39–50.

19 Monteleone P, Maj M: Genetic susceptibility to eating disorders: associated polymorphisms and pharmacogenetic suggestions. Pharmacogenomics 2008;9:1487–1520.

20 National Institute for Clinical Excellence: Eating disorders: core interventions in the treatment and management of anorexia nervosa, bulimia nervosa and related eating disorders. London, NICE, 2004.

21 Kaye W, Gendall K, Strober M: Serotonin neuronal function and selective serotonin reuptake inhibitor treatment in anorexia and bulimia nervosa. Biol Psychiatry 1998;44:825–838.

22 Walsh BT, Kaplan AS, Attia E, Olmsted M, Parides M, Carter JC, Pike KM, Devlin MJ, Woodside B, Roberto CA, Rockert W: Fluoxetine after weight restoration in anorexia nervosa: a randomized controlled trial. JAMA 2006;295:2605–2612.

23 Greenberg BD, Tolliver TJ, Huang SJ, Li Q, Bengel D, Murphy DL: Genetic variation in the serotonin transporter promoter region affects serotonin uptake in human blood platelets. Am J Med Genet 1999;88:83–87.

24 Serretti A, Artioli P: From molecular biology to pharmacogenetics: a review of the literature on antidepressant treatment and suggestions of possible candidate genes. Psychopharmacology (Berl) 2004; 174:490–503.

25 Monteleone P, Santonastaso P, Tortorella A, Favaro A, Fabrazzo M, Castaldo E, Caregaro L, Fuschino A, Maj M: Serotonin transporter polymorphism and potential response to SSRIs in bulimia nervosa. Mol Psychiatry 2005;10:716–718.
26 Erzegovesi S, Riboldi C, Di Bella D, Di Molfetta D, Mapelli F, Negri B, Tenace M, Bellodi L: Bulimia nervosa, 5-HTTLPR polymorphism and treatment response to four SSRIs: a single-blind study. J Clin Psychopharmacol 2004;24:680–682.
27 Steiger H, Joober R, Gauvin L, Bruce KR, Richardson J, Israel M, Anestin AS, Groleau P: Serotonin-system polymorphisms (5-HTTLPR and –1438G/A) and responses of patients with bulimic syndromes to multimodal treatments. J Clin Psychiatry 2008;69:1565–1571.
28 Parsons MJ, D'Souza UM, Arranz MJ, Kerwin RW, Makoff AJ: The –1438A/G polymorphism in the 5-hydroxytryptamine type 2A receptor gene affects promoter activity. Biol Psychiatry 2004;56:406–410.
29 Vandenbergh DJ, Persico AM, Hawkins AL, Griffin CA, Li X, Jabs EW, Uhl GR: Human dopamine transporter gene *(DAT1)* maps to chromosome 5p15.3 and displays a VNTR. Genomics 1992;14: 1104–1106.
30 VanNess SH, Owens MJ, Kilts CD: The variable number of tandem repeats element in DAT1 regulates in vitro dopamine transporter density. BMC Genet 2005;6:55.
31 Davis C, Levitan RD, Kaplan AS, Carter J, Reid C, Curtis C, Patte K, Kennedy JL: Dopamine transporter gene (DAT1) associated with appetite suppression to methylphenidate in a case-control study of binge eating disorder. Neuropsychopharmacology 2007;32:2199–2206.
32 Branson R, Potoczna N, Kral JG, Lentes KU, Hoehe MR, Horber FF: Binge eating as a major phenotype of melanocortin 4 receptor gene mutations. N Engl J Med 2003;348:1096–1103.
33 Potoczna N, Branson R, Kral JG, Piec G, Steffen R, Ricklin T, Hoehe MR, Lentes KU, Horber FF: Gene variants and binge eating as predictors of comorbidity and outcome of treatment in severe obesity. J Gastrointest Surg 2004;8:971–982.

Prof. Palmiero Monteleone, MD
Department of Psychiatry, University of Naples SUN
Largo Madonna delle Grazie
IT–80138 Naples (Italy)
Tel. +39 081 566 6517, Fax +39 081 566 6523, E-Mail monteri@tin.it

Schwab M, Kaschka WP, Spina E (eds): Pharmacogenomics in Psychiatry.
Adv Biol Psychiatry. Basel, Karger, 2010, vol 25, pp 118–134

Future of Personalized Prescription in Psychiatry

Jose de Leon

University of Kentucky Mental Health Research Center at Eastern State Hospital, and University of Kentucky Colleges of Medicine and Pharmacy, Lexington, Ky., USA; Psychiatry and Neurosciences Research Group (CTS-549), Institute of Neurosciences, University of Granada, Granada, Spain

Abstract

Personalized prescription and pharmacogenomics are related concepts, but are not the same. The 'Introduction' describes the concept of pharmacogenomics, which can be included within personalized prescription, and the role of the Human Genome Project and the Food and Drug Administration in promoting advances in these concepts. In the author's comprehensive view of personalized prescription, clinicians need to consider genetic, environmental and personal variables when prescribing any medication. Known important genetic variables in specific drug responses can be explored by pharmacogenetic tests. Environmental variables – such as co-medication, herb supplements, foods, beverages and smoking – may be much more important than genetic factors for some drugs. Personal factors such as age, gender or medical illnesses (renal or hepatic insufficiency) may be crucial personal variables in the response to some other drugs. The pharmacological knowledge needed to understand personalized prescription includes pharmacokinetics and pharmacodynamic actions, efficacy and safety, idiosyncratic and dose-related adverse drug reactions, prescriber's role and therapeutic window, and linear versus non-linear pharmacokinetics. The applications of these pharmacological concepts in psychiatry are briefly reviewed. Risperidone personalized prescription is provided as an example by describing personalized risperidone selection and personalized risperidone dosing. The future of pharmacogenomic tests and personalized prescription in psychiatry is briefly summarized.

Introduction

While the theme of this book is pharmacogenomics in psychiatry, this particular chapter focuses on personalized prescription rather than on pharmacogenomics. Personalized prescription and pharmacogenomics are related concepts, but are not the same. One of the problems in medicine is that many of its concepts are non-classical (or fuzzy) concepts with unclear but overlapping limits. This section reviews the concept of pharmacogenomics, the role of the Human Genome Project, the concepts of personalized medicine and personalized prescription, and the crucial role of the Food and Drug Administration (FDA) in promoting clinical applications. The article then describes the necessary pharmacological knowledge for understanding personalized prescription and its applications in psychiatry, with personalized prescription of risperidone as an example. Finally, the future of pharmacogenomic testing and personalized prescription in psychiatry is summarized.

Pharmacogenomics

Pirmohamed [1] defined pharmacogenomics as the study of all genes in the genome that may determine a drug response. Pharmacogenomics has gained major impetus from technological advances and the Human Genome Project.

Human Genome Project

The end of the 20th century brought new hopes of a revolution in medicine based on our advancing knowledge of the human genome. The Human Genome Project, officially completed in 2000 [2], was a crucial step but led to exaggerated hopes. In an example of this hyperbolic optimism, McKusick [3] published a commentary in *JAMA* comparing the Human Genome Project with the revolutionary influence of a 16th century text on anatomy by Vesalius which led to major developments in medicine, including further applications of the scientific method and the development of other basic sciences.

The Human Genome Project was possible due to rapid advances in genetic technologies that made possible the parallel testing of many single nucleotide polymorphisms (SNP) with progressively lowered costs. Currently, one can test millions of SNPs for less than USD 1,000 per sample, and the price is rapidly decreasing. The onset of these rapid technological advances led to a *Science* editorial comment in 1997 that defined 'personalized prescription' as 'tailoring drugs to a patient's genetic make-up' and predicted that personalized prescription would 'soon' reach clinical practice [4]. More exact estimates for the year in which generalized use of personalized prescription would begin included 2015 according to a 1999 *Time* magazine article [5] and 2020 according to a 2001 *JAMA* article [6].

Recent developments have proven how naïve it was to think that human genome mapping would change medicine by 2015 or 2020. We do not yet know the function of approximately one third of human genes; other types of genetic variations such as deletions or duplications, the so-called copy number variations (CNV), may have been neglected [7]. Unfortunately, many of the current platforms and systems used for genotyping mainly pay attention to SNPs and neglect CNVs or less common genetic variations such as microsatellite polymorphisms and translocations, inversions and substitutions, which may have some pharmacogenomic relevance [7]. Finally, the relevance of epigenetics to human pharmacogenetic response is not well understood

[8], but it is important to know that in one animal model, drug tolerance was caused by epigenetic mechanisms [8].

Since the 'genomic' boom, technological advances have facilitated the development of a new wave of parallel testing of multiple physiological substrates and of new disciplines – including transcriptomics, proteomics and metabolomics [9]. These new technologies, sometimes included under the heading of biomarkers, provide hundreds or thousands of pieces of data on each patient, but produce two types of intrinsic problems: untested statistical analyses and complex interpretation of the results. Regarding statistical problems, the traditional statistical tests were developed to test one or a few hypotheses in dozens or hundreds of patients (many more individuals than tested hypotheses), not to test hundreds and thousands of hypotheses in samples that frequently include fewer individuals than hypotheses. Several new statistical methods are being developed to manage these large statistical databases. The author has collaborated in attempts to use two statistical methods derived from engineering statistics: analyzing genetic data using data mining [10] and a derivation of sensitive analysis and systems engineering [11]. Despite these preliminary attempts, he acknowledges that there are no validated methods; moreover, the statistical method used in genome-wide association studies with thousands of patients and published in the best scientific journals demonstrates very poor replicability.

Results from some of the new tests or biomarkers are hard to interpret since we know little about the normal variations of physiological substrates. In the case of metabolomics, we know little about the normal values of the hundreds of lipids that can be found in human blood. Even if we focus on pharmacogenomics and on the simplest genetic variations, SNPs, we have limited understanding of how to extrapolate to the clinical environment a statistical association between a specific SNP and response to drug X in a well-controlled study. The specific SNP may be associated with functional changes, may not be associated with functional changes and may be linked with other functional SNPs, or may be explained by a false-positive result. The relationships between SNPs and gene function appear to be fairly complicated in some of the well-studied genes. A well-studied pharmacogenetic gene, cytochrome P450 2D6 (CYP2D6), has more than 90 known genetic variations (including SNPs and CNVs) and more than 60 alleles [12]. The functional effects of some of the rarest CYP2D6 alleles are not known despite being relatively easy to study using a phenotyping test that requires giving a pill and measuring urine metabolites. The racial variations are relatively well understood, and it is thought that measuring approximately 20 alleles may provide a reasonable amount of information to clinicians about CYP2D6 phenotypes and function [13]. In most patients, these 20 alleles would establish whether or not the enzyme is present, and, if present, whether it is under- or overactive. Unfortunately, learning the functionality of these genetic variations (the phenotype-genotype association or correlation) has taken over 20 years to develop. Only a few SNPs have been studied in most genes of possible interest in psychiatric pharmacogenomics; in most of them, the functional meaning of these SNPs is not known. The conceptual and scientific difficulties in extrapolating from basic research to clinical applications, usually called translational research, are usually ignored in the literature and may be among the major obstacles for applying pharmacogenomics or personalized prescription in the clinical environment.

Concepts of Personalized Medicine and Personalized Prescription

The concepts of personalized or individualized medicine and prescription are not new in medical parlance. However, genetic advances have made discussing 'personalized medicine' and 'personalized prescription' in genetic terms fashionable. In fact, even lay journals use these concepts frequently, referring mainly to genetic differences between patients. In introducing the first issue of the newly created journal *Personalized Medicine,* Ruaño [14] reminded us that physicians have traditionally practiced personalized medicine in their attempts to decide the best treatment for each of their patients. However, physicians were not using the term 'personalized medicine'; the personalized approach traditionally used by physicians was probably based on subjective physician preferences and not on scientific knowledge. In fact, psychiatrists had used the term in a completely different way. In 1952, Osborn [15] titled his psychiatry textbook *Psychiatry and Medicine: An Introduction to Personalized Medicine*. The idea behind that title was that each patient is a unique individual with unique psychological mechanisms. For Osborn, the principle that allowed personalized medicine was not genetics, but psychoanalysis. Obviously, Osborn's opinion appears somewhat outmoded. On the other hand, the current exclusive focus of personalized medicine on genetics may be wrong.

This author [7] views personalized medicine as a very global concept that may include 'personalized surgery', 'personalized rehabilitation', 'personalized nutrition' and other types of personalized medical interventions and, more importantly for psychiatrists, 'personalized prescription'. Personalized prescription should include not only the use of new tests, which may or may not be pharmacogenetic tests, but also the consideration of all scientific information valid for prescribing medication [16]. Pharmacology is a mechanistic science; knowing the pharmacological principles behind drug response allows predictions to be made (table 1). For many drugs, genetic factors may be irrelevant in drug response or may be much less important than other non-genetic factors. Our pharmacological knowledge of each drug should determine what aspects are important in that drug's personalized prescription. In this comprehensive view of personalized prescription, clinicians need to consider genetic, environmental and personal variables when prescribing any medication [16]. Known important genetic variables in specific drug response can be explored using pharmacogenetic tests. Environmental variables such as co-medication, herb supplements, foods, beverages, and smoking may be much more important than genetic factors for some drugs. Personal factors such as age, gender or medical illnesses (renal or hepatic insufficiency) may be crucial personal variables in the response to some other drugs.

Table 1. Pharmacological principles behind the author's view of personalized prescription

1	Mechanistic science: Pharmacology is a mechanistic science.
2	Pharmacokinetics and pharmacodynamics: Drug response is explained by the pharmacokinetic and pharmacodynamic actions of the drugs. Pharmacokinetic actions are usually a first step, occurring prior to pharmacodynamic actions. Pharmacodynamic actions in the brain may require a greater level of complexity than pharmacodynamic actions in the periphery, due to the complexity of moving drugs from the blood to the brain target.
3	Genetic, environmental and personal variables: Influence drug pharmacokinetics and pharmacodynamics.
4	Pharmacogenetic tests: Any attempt to use pharmacogenomic tests to help with personalized prescription also needs to take into account environmental and personal variables. For some drug responses, genetic variations may be irrelevant or have relatively little influence when compared with environmental and personal variables.
5	Other important pharmacological principles: A pharmacogenomic test or any other method to personalize prescription for a specific drug or group of drugs can only be developed by taking into account those pharmacological principles that are important in explaining how that drug works in the real world. Personalized medicine must consider (in addition to pharmacokinetics and pharmacodynamics): 5.1. Efficacy and safety (safety includes idiosyncratic versus dose-related adverse drug reactions). 5.2. Prescriber's role and therapeutic window (therapeutic window includes linear versus nonlinear pharmacokinetics).

This author has also hypothesized that personalized prescription can be expressed in two main ways: in the clinical environment as personalized selection of the drug, and as personalized dosing [7]. The author's definition of personalized prescription may be original, but his approach is not new since it is based on advances of pharmacological knowledge that are usually expressed in the drug prescribing information required by the FDA and in pharmacological textbooks.

Crucial Role of the FDA in Promoting the Use of Personalized Prescription and Pharmacogenetic Testing

The FDA has had a crucial role in promoting the use of pharmacological information to personalize prescriptions and in the introduction of pharmacogenetic tests in the clinical environment. A remarkable step was the terfenadine story. Terfenadine is a non-sedating antihistaminic that was approved by the FDA in the late 1980s with an average recommended dosage for average subjects after rigorous placebo-controlled clinical trials in healthy subjects not taking other medications. Then, a second experiment, naturalistic and not well-controlled, began when terfenadine was given to the general population which included many non-average subjects who were ill or taking other medications. In 1996, it was clear that more than 100 people had died in the USA in this naturalistic experiment. The 'average' doses were toxic and caused arrhythmias in subjects taking some co-medications inhibiting the cytochrome P450 (CYP) isoenzyme that metabolizes terfenadine. Had our knowledge of CYP 'science' been better, the lethal outcomes during naturalistic use would have been avoided. The deaths of more than 100 people taking terfenadine were not in vain. The FDA began progressively forcing companies to study drug-drug interactions (DDIs) and CYP metabolism. Thus, drug package inserts (currently called prescribing information) were required to include progressively more information on DDIs (environmental variables) and on peculiar situations such as renal or liver insufficiency (personal variables). After terfenadine, several drugs were withdrawn from the market due to similar cases of heart toxicity associated with DDIs, leading to drug prescribing information that increasingly focused on non-average subjects by including information on the effects of environmental and personal variables in drug response.

A further development in the FDA's approach was the inclusion of genetic information in the drug prescribing information. This was not well-received by pharmaceutical companies which had not embraced pharmacogenomic testing in clinical practice. In fact, an FDA official related [17] that when she met with drug industry representatives in 2001 to discuss the promise of personalized medicine: 'People stood up and said: We are terrified'. It is very easy to explain this terror. The pharmaceutical companies' current business model assumes drug approval on the basis of an average dosage recommendation for an average patient. Thus, the practice of excluding some patients from the drug using pharmacogenomic tests would narrow market niches. On the other hand, genotyping and treating some patients with alternative dosages would complicate prescribing information relative to competing drugs [18, 19]. If a drug is approved with pharmacogenomic testing as a requirement, the marketers of previously approved competing drugs would surely remind physicians that their drugs do not have such a requirement in their prescription package, but the new drug does.

The FDA has progressively set new recommendations to promote pharmacogenetics and personalized prescription. In 2005, the FDA provided guidance for the drug industry regarding pharmacogenetic data submission [20] that described a metabolic enzyme important for psychiatry, CYP2D6, as a 'valid biomarker' and introduced the idea of a voluntary data submission program. In 2008, the FDA [21] issued draft guidance for 'in vitro diagnostic multivariate index assays' (IVDMIAs). Pharmacogenomic, metabolomic and proteomic tests are IVDMIAs, and thus the FDA was indicating its intent to require IVDMIAs to meet pre-market and post-market device requirements under FDA regulations. Prior to that, the FDA had not been involved in regulating diagnostic tests.

In addition, the FDA took two major practical steps: in 2006 it approved the first pharmacogenomic test, the AmpliChip CYP 450 Test [18, 19], and in 2007 it began requiring clinicians to use a pharmacogenomic test before administering carbamazepine in a particular racial subgroup [22].

Roche Molecular Systems, Inc., developed the first pharmacogenomic test approved by the FDA, the AmpliChip CYP 450 Test. The microarray contains over

Efficacy	Safety
Pharmacological mechanisms explaining drug response	Pharmacological mechanisms explaining ADRs 1 Dose-dependent ADRs – Explained by an augmentation of the pharmacological action 2 Idiosyncratic ADRs – Not predicted by pharmacological action – Two main mechanisms are suggested: peculiar metabolic pathways leading to reactive metabolites and/or immunological response

Fig. 1. Efficacy and safety. ADRs = Adverse drug reactions.

15,000 oligonucleotide probes allowing testing for 20 CYP2D6 alleles, 7 CYP2D6 duplications, and 3 cytochrome P450 2C19 (CYP2C19) alleles [18, 19]. CYP2D6 is particularly important in psychiatry since it metabolizes many antipsychotics and antidepressants. CYP2D6 poor metabolizers (PM) lack CYP2D6 in their bodies; their prevalence varies by race (highest among Caucasians, approximately 7%, and lower in other races, 1–3%). The psychiatric applications of CYP testing have been described in prior articles [12, 23, 24].

The FDA required that HLA-B*1502 genotyping in Asians should be performed before prescribing carbamazepine to avoid the almost certain development of Stevens Johnson syndrome/toxic epidermal necrolysis in those with this marker [22].

In conclusion, in the view of the author: (1) pharmacogenomic tests are one type of personalized prescription test, and (2) personalized prescription should consider the influence of genetic, environmental and personal variables on each drug.

Pharmacological Knowledge Needed for Personalized Prescription

Table 1 describes drug response as being explained by pharmacokinetic and pharmacodynamic actions influenced by genetic, environmental and personal variables. Pharmacogenetic testing, and any other form of personalized prescription, must focus on other basic pharmacological principles besides pharmacokinetics and pharmacodynamics. This chapter cannot include a pharmacology textbook; therefore, tables and figures are used to remind the reader of crucial pharmacological concepts. Efficacy and safety are important concepts described in figure 1. Within safety, idiosyncratic versus dose-related adverse drug reactions (ADRs) must be distinguished [7]. Similarly, the prescriber's role and therapeutic window or index needs to be considered [16]. Pharmacology is very important in predicting drug response in drugs with a narrow therapeutic index. The prescriber's choice may be more relevant than pharmacological

Linear pharmacokinetics

Concentration increases and decreases in a linear pattern in relation to the dose.
The concentration:dose ratio can be represented by a straight line.

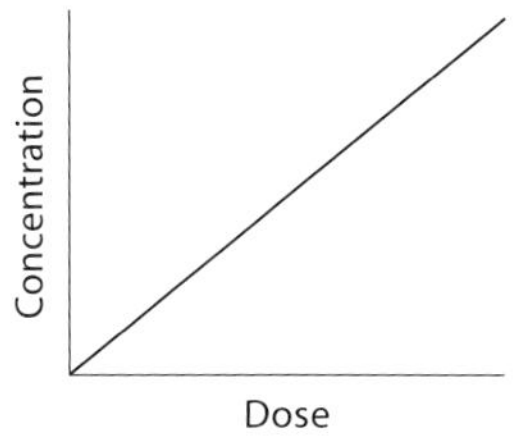

Non-linear pharmacokinetics

The concentration:dose ratio CANNOT be represented by a line making it more difficult to personalize dosing.
Several pharmacological mechanisms can explain this:

- Drugs that are auto-inhibitors of their own metabolism (e.g. paroxetine and fluoxetine)
- Drugs that are auto-inducers of their own metabolism (e.g. carbamazepine)
- Changes influenced by protein-binding and free concentration (e.g. valproate)
- Saturation kinetics: in high concentrations the half-life increases dramatically (e.g. phenytoin)
- Saturable absorption (e.g. gabapentin absorption decreases at higher doses)

Fig. 2. Comparing linear versus non-linear pharmacokinetics. Valproate concentration does not increase proportionally with the dose, but increases to a lesser extent due to saturable plasma protein binding. It is believed that the binding rate is 85–95% at low doses and 70% at high doses. Phenytoin has a narrow therapeutic window and follows non-linear pharmacokinetics, which is dose-dependent and capacity-limited. This means that it is possible to attain excessive drug concentrations with modest dosage increases. When the drug reaches toxic levels phenytoin half-life may increase in a dramatic way due to the saturation of its metabolism. In fact, half-lives as long as 140 h have been described in intoxicated patients. Gabapentin appears to move from the intestine to the blood using a saturable L-amino acid transporter; gabapentin bioavailability is not dose-proportional and decreases at higher doses.

Table 2. Comparing wide and narrow therapeutic window drugs

	Wide	Narrow
Psychiatric drugs	most of the new drugs	most of the old drugs
Most relevant in dosing	prescriber's choice	pharmacology
Safety of dosing	relatively non-toxic	relatively toxic
Efficacy	low concentrations may be ineffective	varies for different individuals
Therapeutic drug monitoring	not well studied	used in clinical practice
Personalized dosing models	explain little of the drug response	explain more of the drug response
Pharmacogenomic tests	poor predictors of dosing	better predictors of dosing

principles in drugs with a wide therapeutic index. Table 2 elaborates on how the therapeutic window may influence personalized dosing. Another pharmacological concept that is important in dosing is linear versus non-linear pharmacokinetics (fig. 2). Pharmacokinetic type is fundamental in establishing personal dosing and in predicting the relationships between drug dose and drug blood concentration. It is easy to predict concentration using dose with drugs displaying linear kinetics, but more complicated with non-linear kinetics, which is present in some psychiatric drugs (fig. 2).

Applying Personalized Prescription in the Clinical Practice of Psychiatry

By now it should be evident to the reader that it is impossible to begin thinking about how to apply personalized prescription in clinical practice without a thorough understanding of the pharmacokinetic and pharmacodynamic principles of psychiatric drugs. Table 3 includes a brief attempt to summarize the pharmacological mechanisms of the most important psychiatric drugs. Table 4 explains the pharmacological principles behind personalized drug selection, and provides psychiatric examples of personalized drug selection in clinical practice. Table 5 explains the pharmacological principles behind personalized dosing, and provides psychiatric examples of personalized drug selection in clinical practice. Personalized drug selection and dosing in psychiatry have been described in more detail in a prior article [7].

Personalized Prescription of Risperidone as an Example

The only way to completely understand what personalized prescription may mean in clinical practice is to provide a good example, such as risperidone. Risperidone prescription may need to take into account genetic, environmental and personal variations. In this article, it is not possible to extensively review the pharmacokinetics and pharmacodynamics of risperidone, which were reviewed in prior articles [25–28] and summarized in tables 6 and 7. Table 6 presents personalized risperidone selection. Basic pharmacological information relevant for risperidone selection is presented in the upper part of the table. Once a clinician has decided that an antipsychotic is needed, the lower part of table 6 describes which factors may be considered for or against risperidone selection. The information in the literature is very limited; multiple pragmatic randomized trials such as the Clinical Antipsychotic Trials of Intervention Effectiveness (CATIE) [29] are needed to correctly inform this decision. Due to the limited information available, and since clinicians frequently have drug preferences and dislikes, different clinicians would likely disagree on whether risperidone is a good first-choice antipsychotic for various patients. The existence of generic and long-acting formulations is a factor in favor of risperidone selection.

Table 7 presents personalized risperidone dosing. Basic pharmacological information relevant for risperidone dosing is presented in the upper part of the table. This table is based more on available scientific and pharmacological principles than the table focused on risperidone selection. As risperidone probably follows linear kinetics in all age groups [28], information on the effects of factors that change risperidone pharmacokinetics in adults may reasonably be extrapolated to children and geriatric patients. Three major factors are considered relevant when adapting risperidone dosing in patients with different personal characteristics: the presence of CYP3A inducers and/or CYP inhibitors, and CYP2D6 PM status (table 6). CYP2D6 PMs lack CYP2D6, the enzyme that has higher affinity for risperidone and may be the most

Table 3. A brief summary of pharmacological mechanisms used to develop personalized prescription in psychiatry

1 Pharmacological principles behind the pharmacodynamics of efficacy are poorly understood and cannot easily be studied in vivo in patients.
   - 1.1 Antipsychotics probably work as D_2 blockers.
   - 1.2 The majority of antidepressants are thought to be serotonin and/or noradrenaline reuptake inhibitors, but we are not sure whether this explains their antidepressant actions.
   - 1.3 It is not known or well understood how antidepressants work in anxiety disorders.
   - 1.4 Benzodiazepines probably act as allosteric modulators of the $GABA_A$ receptors.
   - 1.5 It is not known how the so-called mood stabilizers work.

2 Pharmacological principles behind the brain pharmacodynamics of safety are better understood.
   - 2.1 Increases in appetite secondary to most antipsychotics and some antidepressants may be mainly mediated by H_1 and 5-HT_{2C} blockade.
   - 2.2 Sedation secondary to antipsychotics and some antidepressants may be mainly mediated by H_1, and muscarinic blockade.
   - 2.3 Sedation secondary to benzodiazepines and some mood stabilizers may be mediated by GABAergic actions.
   - 2.4 Extrapyramidal symptoms secondary to antipsychotics are mediated by D_2 blockade.
   - 2.5 Hyperprolactinemia secondary to antipsychotics is mediated by D_2 blockade.
   - 2.6 Cognitive impairment secondary to some antipsychotics and some antidepressants may be mediated by muscarinic blockade.
   - 2.7 The serotonin syndrome secondary to the combinations of antidepressants, lithium and/or other drugs is believed to be explained by increased central and peripheral serotonin release.

3 Peripheral pharmacodynamic mechanisms are not always well studied.
   - 3.1 Changes in heart rate and blood pressure secondary to some antidepressants and some antipsychotics may be mediated by the blockade of α- and β-adrenergic receptors and muscarinic receptors (central components are possible).
   - 3.2 Constipation and urinary retention secondary to some antidepressants and antipsychotics are mainly mediated by muscarinic blockade.
   - 3.3 Sexual ADRs and urinary incontinence secondary to some antipsychotics may be mediated by adrenergic blockade.
   - 3.4 Mechanisms of peripheral glucose and lipid metabolism disturbed by antipsychotics are not well understood. Peripheral glucose and lipid metabolism actions of antidepressants and mood stabilizers are not well studied.
   - 3.5 Platelet, gastrointestinal and sexual ADRs from some antidepressants are probably mediated by peripheral serotonergic changes.
   - 3.6 Cardiac ion channel actions produced by some antidepressants and antipsychotics are not well studied. How these actions contribute to increased sudden deaths is not well understood.
   - 3.7 Potentially lethal ADRs secondary to antipsychotics and mood stabilizers in skin, liver and hematological tissues are not understood and in some cases are possibly immunologically mediated.

4 Pharmacokinetic mechanisms are better understood but tend to be ignored by psychiatrists.
   - 4.1 CYP.
     - CYP2D6 is important for some antipsychotics, some antidepressants and activation of some opioids.
     - CYP3A is important for some antipsychotics, some antidepressants, carbamazepine and some benzodiazepines.
     - CYP2C19 is important for some antidepressants and diazepam.
     - CYP1A2 is important for clozapine and olanzapine.
   - 4.2 UGT. These poorly understood enzymes are important for lamotrigine, valproate, some opioids and secondary pathways for some antipsychotics and possibly some antidepressants.
   - 4.3 Transporters are poorly understood.
     - P-glycoprotein may be important at the intestine, liver and blood-brain barrier.
     - Kidney transporters may be important for lithium and other drugs.

Table 4. Pharmacological principles behind personalized drug selection and examples in psychiatry

1 Complexity: Personalizing drug selection is complicated since it requires considering multiple drugs.
2 Risk/benefit: The risk and benefits of each drug need to be considered.
3 Indication: The level of complexity is much higher when drugs from different classes can be selected (e.g. mania, which includes mood stabilizers and antipsychotics) versus only one class (e.g. a non-affective psychotic episode, which only includes antipsychotics).
4 Personal opinions: Personal opinions of physicians and patients on particular drugs and ADRs may be important in personalizing drug selection and are hard to predict using statistical models.

Examples of personalized drug selection in clinical practice[1]

1 Contraindication of some drugs in some unusual subjects:
   - Genetic variation. HLA-B*1502 genotyping in Asians: contraindicates carbamazepine.
   - Environmental variation. Taking drugs that increase QTc: contraindicates ziprasidone.
   - Personal variation. Pregnancy: contraindicates valproate.

2 Exclusion of some drugs in some patients due to some frequent ADRs not common to all drugs in one class:
   - Genetic variations. Variants associated with TD risk: recommend against some antipsychotics.
   - Environmental variations. Prescription of drugs associated with obesity: recommend against antipsychotics with greater risk of causing obesity.
   - Personal variations. Elderly females are prone to develop TD: recommend against antipsychotics associated with greater TD risk.

3 Best drug for each patient:
This is an elusive goal beyond our current knowledge.

TD = Tardive dyskinesia.
[1] For more details, see de Leon [7].

important metabolic pathway for risperidone by hydroxylating it to 9-hydroxyrisperidone. PM patients can be identified by CYP2D6 genotyping [18, 19] or by measuring risperidone trough levels in steady state [23]. If the patient has a plasma risperidone/9-hydroxyrisperidone concentration ratio >1 and is not taking a CYP2D6 inhibitor (bupropion, fluoxetine or paroxetine) then he/she is likely to be a CYP2D6 PM. In the absence of these drugs, if the risperidone/9-hydroxyrisperidone ratio is 2 or 3, the patient is most definitively a CYP2D6 PM. Prior articles [23, 25–28] described this in detail. One of the articles described a pharmacokinetic model including genetic and environmental variables that allowed an exploration of their effects in dosing [26].

Future of Pharmacogenomic Testing in Psychiatry

The only FDA-required pharmacogenomic test in psychiatry is a test for one drug and for one racial group [22], and it only eliminates the risk of a relatively rare idiosyncratic ADR (HLA-B*1502 genotyping in Asians for carbamazepine). Clinicians have complained to the author that having one pharmacogenomic test for drug selection in one race is a miniscule advance. Unfortunately, this is the only pharmacogenomic test in the immediate future of psychiatry that has definitive support for its clinical indication.

Table 5. Pharmacological principles behind personalized dosing and examples in psychiatry

1 Pharmacological principles behind dosing are much simpler than those behind drug selection.

2 Bottle-neck situations. One needs to remember that bottle-neck situations may apply at crucial points in drug response.
- Absorption disturbances by food[1], other drugs[2], or even anatomical problems[3] may decrease a drug's availability and render a drug ineffective.
- It is not well understood whether difficulties in crossing the BBB may be relevant in psychotropic drugs. Problems at the BBB have been explored as a possible mechanism for explaining lack of anticonvulsant response. Differences in BBB permeability may explain differences in toxicity between risperidone and its metabolite, 9-hydroxyrisperidone, marketed as paliperidone.

3 Dosing models combining genetic, environmental and personal variables may focus on:
- 3.1 Pharmacokinetic models: Models predicting blood drug concentration. Developing models for drugs following linear kinetics is much easier than for those drugs following non-linear kinetics (table 4). The clinical relevance of these models depends on how well blood concentration predicts drug response (efficacy and safety).
- 3.2 Clinical models: Predicting ADRs. They should incorporate pharmacokinetic and pharmacodynamic information. The ADR type is important:
  - 3.2.1 Idiosyncratic ADRs: It makes no sense to try to develop dosing models since they are not dose related. Idiosyncratic ADRs may need certain doses (probably low) and after that the dose is irrelevant. Thus, it is better to deal with these ADRs by personalizing drug selection.
  - 3.2.2 Dose-dependent ADRs: Dosing models may have better potential. However, it is important to consider whether tolerance develops or not.
    - 3.2.2.1 ADRs not subject to tolerance: It may be easier to develop predictive dosing models. The narrower the therapeutic window, the more likely that the predictive model may work in the clinical environment.
    - 3.2.2.2 ADRs subject to tolerance: High dosing may predict toxicity only in the initial doses. After some initial doses the patient may develop tolerance to ADRs. The predictive model for dosing needs to consider the duration of the treatment (initial versus maintenance).

4 The role of therapeutic window.
- 4.1 Wide therapeutic window drugs:
  - Genetic variations: Being a UM may explain lack of efficacy.
  - Environmental variations: Taking inducers may explain lack of efficacy.
- 4.2 Narrow therapeutic window drugs:
  - Genetic variations:
    Being a UM may explain lack of efficacy when the drug is metabolized.
    Being a PM may explain lack of efficacy when the drug is activated.
    Being a UM may explain ADRs when the drug is activated.
    Being a PM may explain ADRs when the drug is metabolized.
  - Environmental variations:
    Taking inducers may explain lack of efficacy when the drug is metabolized.
    Taking inhibitors may explain lack of efficacy when the drug is activated.
    Taking inhibitors may explain ADRs when the drug is metabolized.
  - Personal variations: Renal insufficiency may explain ADRs when the drug is excreted by the kidney.

Examples in clinical practice of personalized drug dosing[4]
- Genetic variations: CYP2D6 PMs need half the usual dosages of TCAs.
- Environmental variations: Taking fluvoxamine requires lower clozapine dosing.
- Personal variations: Advanced age requires decreasing risperidone dosing.

BBB = Blood-brain barrier; PM = poor metabolizer; UM = ultrarapid metabolizer.
[1] Ziprasidone needs to be administered with food to increase its absorption.
[2] Carbamazepine suspension should not be administered with chlorpromazine suspension since they may lead to a precipitate and loss of absorption.
[3] Paliperidone capsules should be avoided in individuals with gastrointestinal narrowing which may hinder passage of a capsule through the gastrointestinal tract.
[4] For more details, see de Leon [7].

Table 6. Personalizing risperidone selection

Pharmacokinetic and pharmacodynamic profiles to consider

1 Important pharmacokinetic differences (which can be handled by dosing, see table 7)

- Risperidone versus non-dependent CYP2D6 antipsychotics (e.g. clozapine, olanzapine, paliperidone, quetiapine and ziprasidone): risk of overdosing in CYP2D6 PMs. This can be corrected by dosing.
- Clinically relevant DDI with CYP inhibitors (relatively similar to other antipsychotics except for ziprasidone and paliperidone, which may be relatively free of them). This can be corrected by dosing.
- Clinically relevant DDI with inducers (relatively similar to other antipsychotics except for ziprasidone, which may be relatively free of them). This can be corrected by dosing.
- Risperidone versus antipsychotics with limited renal excretion (e.g. clozapine, olanzapine, quetiapine and ziprasidone): risk of overdosing patients with renal insufficiency. This can be corrected by dosing.

2 Safety profile

- Increasing QTc: low risk (better than first-generation APs and ziprasidone).
- EPS: intermediate risk (better than first-generation APs and worse than other second-generation APs).
- Metabolic ADRs (obesity, hyperglycemia and hyperlipidemia): intermediate.
- Hyperprolactinemia: high risk (similar to first-generation APs and worse than other second generation APs).
- Sedation: low risk (better than clozapine, olanzapine, phenothiazines and quetiapine).
- Orthostatic hypotension: high risk (probably similar to clozapine, phenothiazines, quetiapine and ziprasidone).
- Sexourinary symptoms: rare (but probably higher than other second-generation APs).
- GI symptoms: lower than the worst APs (aripiprazole and ziprasidone).
- Seizure: probably average risk (lower than the worst APs: chlorpromazine, clozapine, olanzapine and quetiapine).
- Liver toxicity: very low risk (lower than the worst APs: clozapine, olanzapine and phenothiazines).
- Antimuscarinic ADRs: no risk or very low risk (versus those APs with high risk: clozapine, olanzapine, phenothiazines and quetiapine). Remember that due to the potential for EPS, anticholinergic drugs may be needed when taking risperidone.

Personalizing risperidone selection within antipsychotics

1 Risperidone is contraindicated by:

- History of prior severe ADRs on risperidone (and possibly on paliperidone).
- Parkinson's disease and other neurodegenerative illnesses with parkinsonian symptoms. Use aripiprazole, clozapine or quetiapine.
- Current or past prolactin-sensitive cancer. If an AP is really needed, use an AP not associated with relevant prolactin elevations (aripiprazole is better or clozapine).

2 Risperidone may not be a good choice (other APs may be better):

- When orthostatic changes are a risk (serious cardiovascular disease or treatment with antihypertensives): Better choices may be aripiprazole, haloperidol, olanzapine or paliperidone.
- History of EPS in first-generation antipsychotics. Better choices are any of the other second-generation APs except paliperidone.
- Metabolic syndrome: risperidone has an intermediate position (worse than aripiprazole, haloperidol, molindone or ziprasidone; better than clozapine, olanzapine, phenothiazines and quetiapine).

3 Positive aspects of risperidone formulation:

- Generic is available (cheaper than other second-generation APs, and more expensive than first-generation APs).
- Long-acting is available (only others available in USA are fluphenazine, haloperidol and paliperidone).

APs = Antipsychotics; EPS = extrapyramidal symptoms.

Table 7. Personalizing risperidone dosing

	Genetics	Environment	Personal variables
Pharmacokinetic and pharmacodynamic profiles to consider			
CYP2D6	PM: absent	inhibitors: BUP, PAR, FLU[1]	
CYP3A		CYP3A inhibitors CYP3A inducers: CBM[2]	
Renal excretion			renal insufficiency aging: decreased clearance
Volume of distribution			children: lower
D_2 receptors[1]		prior APs: tolerance	geriatric: ↓ demented: ↓↓ adults with MR: ↓

	Normal	Inducers[2] (multiply by 2)	Inhibitors[3,4] (divide by 1.3)	CYP2D6 PM or FLU[4] (divide by 2)
Personalizing risperidone dosing, mg/day Lower doses may be needed when the patient is at risk of an ADR (due to co-medications with pharmacodynamic interactions or illnesses)				
Starting dosage				
Average adult	1–2	2–4	0.75–1.5	0.5–1
First episode	1	2	0.75	0.5
Adult MR	1–2	2–4	0.75–1.5	0.5–1
Geriatric (or RI)	0.25	0.5	0.12	0.12
Demented	0.5	1	0.25	0.25
Children 15–20 kg	0.25	0.5	0.12	0.12
Children ≥20 kg	0.50	1	0.25	0.25
Target dosage				
Average adult	4	8	3	2
First episode	2	4	1.5	1
Adult MR	2	4	1.5	1
Geriatric (or RI)	2	4	1.5	1
Demented	1	2	0.75	0.5
Children 15–20 kg	0.5	1	0.25	0.25
Children ≥20 kg	1	2	0.75	0.5
Usual maximum recommended dosage[5]				
Average adult	6	12	4	3
First episode	4	8	3	2
Adult MR	4	8	3	2
Geriatric (or RI)	4	8	3	2
Demented	1.5	3	1	0.75
Children 15–20 kg	no agreement in the literature			
Children ≥20 kg	no agreement in the literature			

Information presented on personalizing risperidone dosing is a major modification of a prior table [28]. BUP = Bupropion; PAR = paroxetine; FLU = fluoxetine; CBM = carbamazepine; APs = antipsychotics; MR = mental retardation; RI = renal insufficiency; ↓ = decrease in receptors; ↓↓ = very important decrease in receptors.

[1] de Leon et al. [27].

[2] The information on CYP3A inducers is based mainly on CBM data. Other CYP3A inducers have not been well studied. Clinically relevant inducers that may have effects similar to CBM are rifampin, phenobarbital, primidone, phenytoin, non-nucleoside reverse transcript inhibitors (efavirenz and delaviridene), dexamethasone, prednisone and St. John's wort.

[3] High doses of sertraline may also be a CYP2D6 inhibitor. FLU is a CYP2D6 and CYP3A inhibitor and may block both risperidone metabolic pathways. It should be considered as a particularly hazardous inhibitor.

[4] The clinically relevant CYP3A inhibitors include fluvoxamine, cimetidine, ketoconazole, erythromycin, clarithromycin, protease inhibitors, grapefruit juice and dilitiazem.

[5] Risperidone up to 16 mg/day (maximum dose) was approved in the USA. At that time, the antipsychotic doses used were too large.

Currently, CYP2D6 and CYP2C19 genotyping appear to have little future. Pharmaceutical companies are eliminating drugs metabolized by CYP2D6 from their pipeline [18, 19]. First-generation antipsychotics tend to be CYP2D6 drugs and may be as efficacious as the new ones. As they are much cheaper, marketing the use of first-generation antipsychotics plus personalized tests, including CYP2D6 genotyping, may be the way to go [11]. Unfortunately, this idea is contrary to the current marketing strategies of the pharmaceutical companies which promote second-generation antipsychotics. Other psychiatric pharmacogenomic tests for clozapine efficacy, clozapine-induced agranulocytosis and antipsychotic-induced metabolic syndrome have been described in prior articles [9, 24].

The author believes that in the near future, in psychiatry, pharmacogenetic tests or other types of complex biomarkers have some potential in two areas [7]: (1) excluding the use of some drugs for some unusual patients (has major potential since neurology provided the first pharmacogenetic test for carbamazepine), and (2) personalizing drug dosing by using pharmacokinetic genes in narrow therapeutic window drugs (has some potential, but these drugs may be irrelevant for clinical practice unless the old antipsychotics are returned to use). There is dubious potential for: (1) selecting some drugs within a class due to ADR or efficacy profile, and (2) selecting dosing in a wide therapeutic window drug. The author does not see short-term potential in finding the best drug for each patient. This 'very sophisticated' level of personalized prescription is beyond our current knowledge and study methodologies [7].

Future of Personalized Prescription in Psychiatry

This author defines a new way of looking at personalized prescription, describing it as the use of genetic, environmental or personal information for selecting drugs and/or prescribing dosages. With this broad definition, personalized prescription can be utilized without waiting for new developments in pharmacogenomic or other biomarker testing. Personalized prescription requires only that sophisticated clinicians understand that genetic, environmental or personal variables influence pharmacokinetic and pharmacodynamic response; the therapeutic window of the drug may also be important. Blood levels, currently called therapeutic drug monitoring, have been used by psychiatrists to personalize dosing for lithium, tricyclic antidepressants and some antipsychotics including clozapine. Unfortunately, all of these are old drugs rarely used by young prescribers in psychiatry. New drug marketing has convinced psychiatrists that they do not need to use these old drugs; thus, using therapeutic drug monitoring in psychiatry appears irrelevant. It also makes teaching this broad view of personalized prescription difficult.

Conclusions

The 'Introduction' describes the concept of pharmacogenomics that can be included within personalized prescription and the role of the Human Genome Project and the FDA in promoting the advancement of these concepts. Personalized prescription and pharmacogenomics are related concepts, but are not the same. In the author's comprehensive view of personalized prescription, clinicians need to consider genetic, environmental and personal variables when prescribing any medication. The pharmacological knowledge needed to understand personalized prescription and its applications in psychiatry includes pharmacokinetic and pharmacodynamic actions, efficacy and safety, idiosyncratic and dose-related ADRs, prescriber's role and therapeutic window, and linear versus non-linear pharmacokinetics. Risperidone personalized prescription is provided as an example by describing personalized risperidone selection and personalized risperidone dosing. The future of pharmacogenomic tests and personalized prescription in psychiatry is briefly summarized.

Acknowledgments

The author thanks Lorraine Maw, MA, and Margaret T. Susce, RN, MLT, for editorial assistance.

Disclosure and Conflict of Interest

Since 10/1/2008, J.d.L. has been a co-investigator in an NIH Small Business Innovation Research Grant awarded to Genomas, Inc. He has received researcher-initiated grants from Eli Lilly (one ended in 2003 and the other, as co-investigator, ended in 2007) and from Roche Molecular Systems, Inc. (ended in 2007). He has been on the advisory board of Bristol-Myers Squibb (2003/04), AstraZeneca (2003) and Roche Molecular Systems, Inc. (2006). His lectures have been supported once by Sandoz (1997), twice by Lundbeck (1999 and 1999), twice by Pfizer (2001 and 2001), three times by Eli Lilly (2003, 2006, and 2006), twice by Janssen (2000 and 2006), once by Bristol-Myers Squibb (2006), and eight times by Roche Molecular Systems, Inc. (twice in 2005 and six times in 2006).

References

1 Pirmohamed M: Pharmacogenetics and pharmacogenomics. J Clin Pharmacol 2001;52:345–347.
2 Golden F, Lemonick MD: The race is over. Time, 2000, pp 18–23.
3 McKusick VA: The anatomy of the human genome: a neo-vesalian basis for medicine in the 21st century. JAMA 2001;286:2289–2295.
4 New research horizons. Science 1997;278:2039.
5 Lertola J: Deciphering the code and what might come from it. Time, 1999, pp 68–69.
6 Collins FS, McKusick, VA: Implications of the human genome project for medical science. JAMA 2001;285:540–544.
7 de Leon J: The future (or lack of future) of personalized prescription in psychiatry. Pharmacol Res 2009;59:81–89.
8 Nebert DW, Zhang G, Vesell ES: From human genetics and genomics to pharmacogenetics and pharmacogenomics: past lessons, future directions. Drug Metab Rev 2008;40:187–224.

9 de Leon J: Pharmacogenomics: the promise of personalized medicine for CNS disorders. Neuropsychopharmacology 2009;34:159–172.

10 Baca-Garcia E, Vaquero-Lorenzo C, Perez-Rodriguez MM, Gratacòs M, Bayés M, Santiago-Mozos R, Leiva-Murillo JM, de Prado-Cumplido M, Artes-Rodriguez A, Ceverino A, Diaz-Sastre C, Fernandez-Navarro P, Costas J, Fernandez-Piqueras J, Diaz-Hernandez M, de Leon J, Baca-Baldomero E, Saiz-Ruiz J, Mann JJ, Parsey RV, Carracedo A, Estivill X, Oquendo MA: Nucleotide variation in central nervous system genes among male suicide attempters. Am J Med Genet B Neuropsychiatr Genet 2010;153B:208–213.

11 Ruaño G, Goethe JW, Caley C, Woolley S, Holford TR, Kocherla M, Windemuth A, de Leon J: Physiogenomic comparison of weight profiles of olanzapine- and risperidone-treated patients. Mol Psychiatry 2007;12:474–482.

12 de Leon J, Susce MT, Johnson M, Hardin M, Maw L, Shao A, Allen AC, Chiafari FA, Hillman G, Nikoloff DM: DNA microarray technology in the clinical environment: the AmpliChip CYP 450 Test for CYP2D6 and CYP2C19 genotyping. CNS Spectr 2009;14:19–34.

13 Cai WM, Nikoloff DM, Pan RM, de Leon J, Fanti P, Fairchild M, Koch WH, Wedlund PJ: CYP2D6 genetic variations in healthy adults in psychiatric African-American subjects: implications for clinical practice and genetic testing. Pharmacogenom J 2006;6:343–350.

14 Ruaño G: Quo vadis personalized medicine. Personalized Med 2004;1:1–7.

15 Osborn LA: Psychiatry and Medicine: An Introduction to Personalized Medicine. New York, McGraw Hill, 1952.

16 de Leon J: The crucial role of the therapeutic window in understanding the clinical relevance of the poor versus the ultrarapid metabolizer phenotypes in subjects taking drugs metabolized by CYP2D6 and CYP2C19. J Clin Psychopharmacol 2007;27: 241–245.

17 Cappell K, Arndt M, Carey J: Drugs get smart. Business Week, 2005, pp 76–85.

18 de Leon J, Susce MT, Murray-Carmichael E: The AmpliChip CYP450 Genotyping Test: integrating a new clinical tool. Mol Diagn Ther 2006;10:135–151.

19 de Leon J: The AmpliChip CYP450 Test: personalized medicine has arrived in psychiatry. Expert Rev Mol Diagn 2006;6:277–286.

20 FDA: Guidance for industry: pharmacogenomic data submission. 2005. www.fda.gov/downloads/RegulatoryInformation/Guidances/ucm126957.pdf.

21 FDA: In Vitro Diagnostic Multivariate Index Assays (IVDMIAs). 2008.

22 FDA: Information for healthcare professionals: carbamazepine (marketed as Carbatrol, Equetro, Tegretol, and generics). 2007.

23 de Leon J, Armstrong SC, Cozza KL: Clinical guidelines for psychiatrists for the use of pharmacogenetic testing for CYP450 2D6 and CYP450 2C19. Psychosomatics 2006;47:75–85.

24 de Leon J, Arranz MJ, Ruaño G: Products for pharmacogenetic testing in psychiatry: a review of features and clinical realities. Clin Lab Med 2008;28: 599–617.

25 de Leon J, Susce MT, Pan RM, Fairchild M, Koch W, Wedlund PJ: The CYP2D6 poor metabolizer phenotype may be associated with risperidone adverse drug reactions and discontinuation. J Clin Psychiatry 2005;66:15–27.

26 de Leon J, Susce MT, Pan RM, Wedlund PJ, Orrego ML, Diaz FJ: A study of genetic (CYP2D6 and ABCB1) and environmental (drug inhibitors and inducers) variables that may influence plasma risperidone levels. Pharmacopsychiatry 2007;40:93–102.

27 de Leon J, Armstrong SC, Cozza KL: A preliminary attempt to personalize risperidone dosing using drug-drug interactions and genetics. Part I. Psychosomatics 2008;49:258–270.

28 de Leon J, Armstrong SC, Cozza KL: A preliminary attempt to personalize risperidone dosing using drug-drug interactions and genetics. Part II. Psychosomatics 2008;49:347–361.

29 Lieberman JA, Stroup TS, McEvoy JP, Swartz MS, Rosenheck RA, Perkins DO, Keefe RSE, Davis SM, Davis CE, Lebowitz BD, Severe J, Hsaio JK: Effectiveness of antipsychotic drugs in patients with chronic schizophrenia. N Engl J Med 2005;353: 1209–1223.

Jose de Leon, MD
Mental Health Research Center at Eastern State Hospital
627 West Fourth St.
Lexington, KY 40508 (USA)
Tel. +1 859 246 7563, Fax +1 859 246 7019, E-Mail jdeleon@uky.edu

Author Index

Subject Index